Invention and Understanding

Invention and Understanding

A Pedagogical Guide to
Three Dimensions

Steven Careau

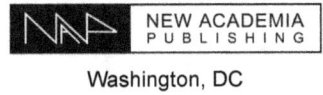

Washington, DC

New Academia Publishing, 2013

Printed in the United States of America

Library of Congress Control Number: 2013949962
ISBN 978-0-9886376-9-6 paperback (alk. paper)

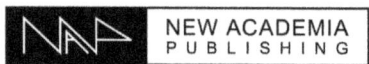 NEW ACADEMIA
PUBLISHING

P.O. Box 27420, Washington, DC, 30028-7420
info@newacademia.com - www.newacademia.com

Portions of the text of Part I have been adapted from "A Pedagogy for Understanding the Visual Arts," in the *Community College Enterprise* 14, no. 2 (Fall 2008): 7–21, and are reprinted here with permission.

For Rachel

Contents

Introduction

We live in a world of objects, and these objects, whether made by human hands or formed through natural processes, possess physical characteristics. Existing in time and changing over time, stationary or moving, all objects have dimensions, occupy space, and respond to gravity. They can be smooth or rough, light or heavy, fragmented or solid. Humans understand this world intuitively. We do not need to actively think about our physical world as we navigate it; we simply "know" it. Our shared childhood experience of the physical environment forms the foundation of our mutual understanding. We all know what a smooth surface feels like. We all understand the dread of falling and the stability of lying down. We all feel the power of a large object when we stand next to it. We know few and many, up and down, here and there. These perceptions are our shared heritage, a heritage that we bring to our encounter with all objects, including works of art.

One of the principal goals of this book is to help teachers guide students to a better understanding of works of art. Though the primary focus of the book is three-dimensional art—objects—essential aspects of the method can be applied to other areas, including disciplines outside the fine arts. Fundamental to the approach are observation and thinking—observing physical reality, and thinking about what is observed. Questions, simply and directly stated, structure this dynamic, serving to illuminate aspects of the physical world that are often intuitively felt but not consciously understood. These questions, dealing with such concerns as material, process, place, and time, require care in answering. The quest is to understand what we are observing, and this questioning, by actively engaging the rational mind, serves to unlock the information that will ultimately help us to grasp the essential nature of the experience. Though the questions deal primarily with the physical aspects of visual reality, issues of culture, of history, context, and theory, inevitably arise in discussion. Thus culture, the individual, and the physical world join in dynamic interrelationship as we strive to understand what we see.

In the first part of the book, in order to answer the question, What do you see? students will follow and ultimately learn a systematic approach to investigating both known objects and challenging works of art. By following a step-by-step approach, students will gain not only insight and confidence but a method of exploration that will serve them as they develop as thinkers able to pursue logical sequences based on available evidence. It is hoped that the book creates a calm yet probing textual environment, one that helps stabilize the inquiry and aids in the development of a thoughtful reflectiveness in students.

In the second part of the book, the same structure of questioning will be used to direct inventive thinking and object making. The reader will be guided through a series of projects that use the questions to generate creative options. While recognizing the role of intuition, this sequencing of options will favor reason over intuition. But the method described in this book does not deny the role of intuition. Working from the perspective that intuitive responses are often guided by past experiences and expectations, this method seeks to enrich and refine intuitive responses by expanding and deepening the foundation upon which creative decisions are made. Guided by an ethos of exploration, students will gain intellectual control of a dynamic often beset with misconceptions and prejudices. This intellectual control will serve our goal of creating adventurous problem solvers and inventive thinkers.

There is a history to the development of this book. In 2001, I published an essay titled "Toward the Unnamed: Creative Strategies in a Fine Arts Seminar" in the *Michigan Community College Journal*. That essay described a strategy of invention that I had developed for an honors seminar. Students in the seminar were guided through a series of projects and exercises that stressed the direct manipulation of materials. The goal was for each student to create a new visual reality that was free of conceptual predetermination. Then, in 2008, I published a second essay, "A Pedagogy for Understanding the Visual Arts," in the *Community College Enterprise*. That essay followed a group of honors students as they learned a systematic approach to understanding works of art. This approach allowed students to gain intellectual control when encountering objects that at first seemed difficult or perplexing.

Invention and understanding: these are the areas of my interest and lie at the heart of the present text. This book is intended both as a pedagogical guide for teachers in the undergraduate classroom and as a study of what was asked of and accomplished by students as they faced the challenges of a demanding creative arts course. It is hoped that the book will give guidance to educators as they, too, face the equally demanding challenges of the contemporary classroom.

PART I

The Questions

The Questions

Do You Recognize It?

We can imagine all works of art falling along a continuum of recognition. At one end of the continuum is the familiar, the known; at the other extreme is the new, the strange. When we look at figure 1.1, for instance, we immediately recognize a human figure—the figure of a female dancer. We have no doubt; recognition is immediate and seemingly automatic. Even if we were to abstract the figure so that only a hint of "figure" was present, the viewer would most likely still see "figure." Our minds recognize shapes and references quickly, even when very little visual information is provided. But when we are confronted with something new, something that we do not readily recognize, as in figure 1.2, we must come to an understanding of its nature without the anchor of recognition. Strangeness often bothers us exactly for this reason: we don't recognize what we see.

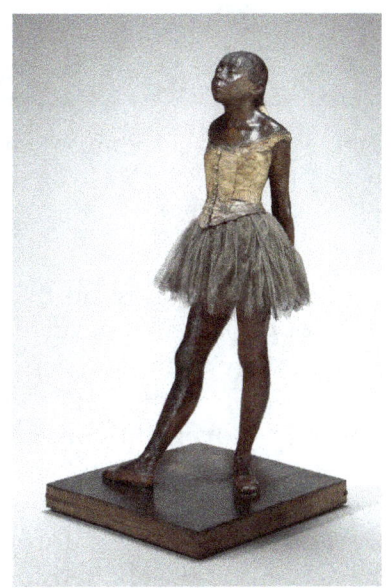

Figure 1.1. Edgar Degas, *Little Dancer Aged Fourteen,* 1878–1881. Yellow wax, hair, ribbon, linen bodice, satin shoes, muslin tutu, wood base, overall without base 38 15/16 × 13 11/16 × 13⅞ in. Courtesy National Gallery of Art, Washington.

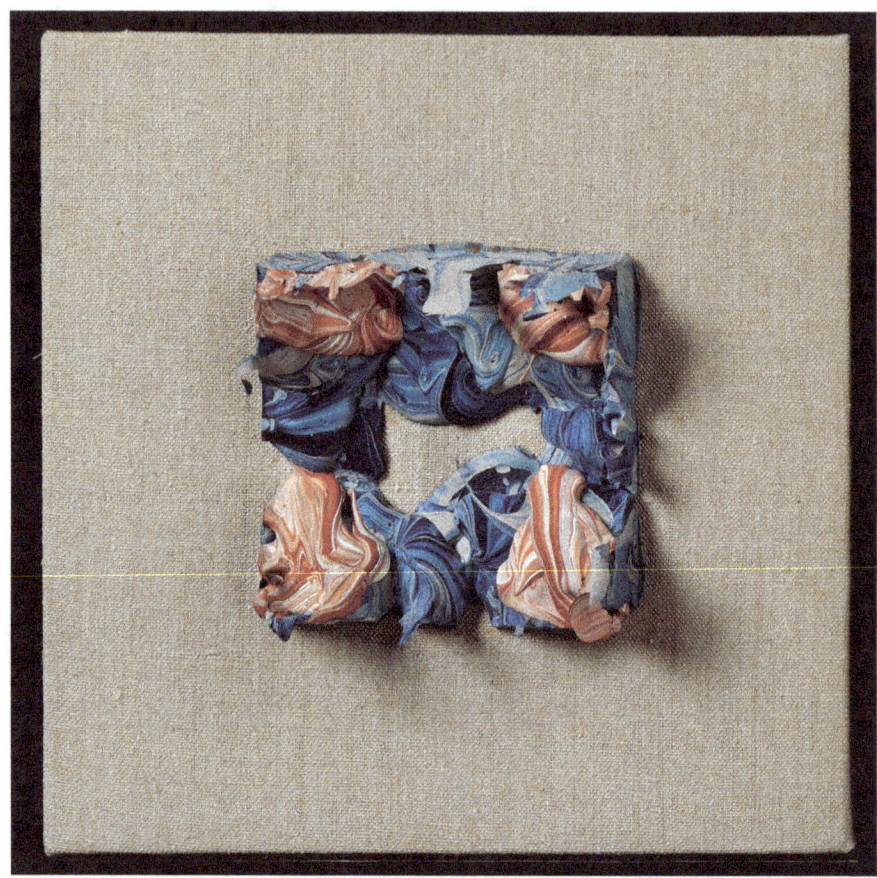

Figure 1.2. A. H. Thompson, untitled, 1986. Thickened and marbleized acrylic paint, 4 × 4 × 1½ in., mounted on linen (shown without integrated black wood frame).

So the first question that we ask is, Do you recognize it? The "it" could be a shape, such as a human figure, or perhaps a style or a strategy used by a particular artist or in a particular tradition. For example, seeing an even repetition of forms might remind us of the art movement minimalism; bizarre imagery and strange juxtapositions, surrealism. The references, at times, seem almost inexhaustible, especially as our knowledge of forms and traditions develops. Some artists, through the uniqueness of their invention, have become closely tied to some shape, gesture, strategy, or material: Kazimir Malevich, geometric imagery; Jackson Pollock, the drip; Richard Serra, curved steel plate. These artists and their innovations have now become part of our cultural inheritance and have taken their place along our continuum of recognition.

The capacity for recognition is one tool for gaining understanding of what we see. But once we have dealt with the question of recognition, we actively move beyond easy classification or bewilderment. Then investigation actually starts.

How Big Is It?

We all understand and respond to size, our bodies serving as innate and lasting instruments of measure. Since our first moments, we have navigated an ever-broadening environment of objects and spaces and have measured them in relationship to our bodies (fig. 1.3). From the moment our parents first lifted us into their arms, we have experienced a relational world of "larger than" and "smaller than." With that knowledge, we have come to understand that size is related to power. The "larger" demands respect and may

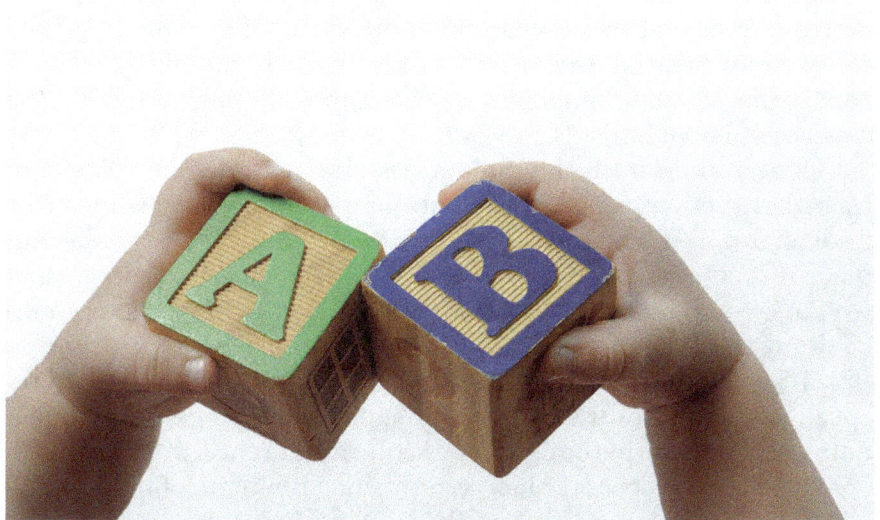

Figure 1.3. Child holding blocks. Photograph by Andy Wainwright, 2012.

evoke feelings of confidence or awe or even intimidation, while the "smaller," allowing greater control, may seem more personal and evoke feelings of intimacy. With time and education, we learn the systems of measurement—of inches and feet, kilometers and cubits—and come to see the world in another way, in a nonintuitive fashion. Though we don't lose the capabilities we developed earlier in life, we learn to describe aspects of the physical world in a more formal fashion, so that we are able to describe an object sitting on a table, for example, as four inches high and quite easy to pick up and use.

Where Is It?

A dialogue exists between object and place, each speaking the language of its identity. The conversation can be loud, with dramatic contrasts, or quiet, marked by subtle differences. Consider a place: a hardwood forest. It is summer, and trees, bushes, and flowers fill our view, surrounding us in all directions. Above the canopy of trees is the sky; through the air move birds and insects. The atmosphere is warm and humid, the odors multiple and changing as we move through the forest. Sounds of all sorts fill and occasionally pierce the air. Let us now place an object, a cube of metal, in the forest, among the tall trees (fig. 1.4). Listen to its voice, its identity— hard, solid, almost changeless, its shape proclaiming geometry, the concept of its making.

Here in the forest, the dialogue is clear and distinct. The contrasts between object and place are marked, and we instinctively understand differences of identity. If we were to take the cube and place it on a pedestal within a white-walled room—the room that we have come to know as a gallery space—the differences would be dramatically lessened; in fact, the cube might seem to fit right in (fig. 1.5). Differences might still exist, but object and place would speak something of the same language, the language of organization, clarity, and, perhaps, thoughtful indifference to the viewer. So if we view object and place in terms of a dialogue of their differences, we come to understand the power of place to create a tone ranging from agreement to disharmony.

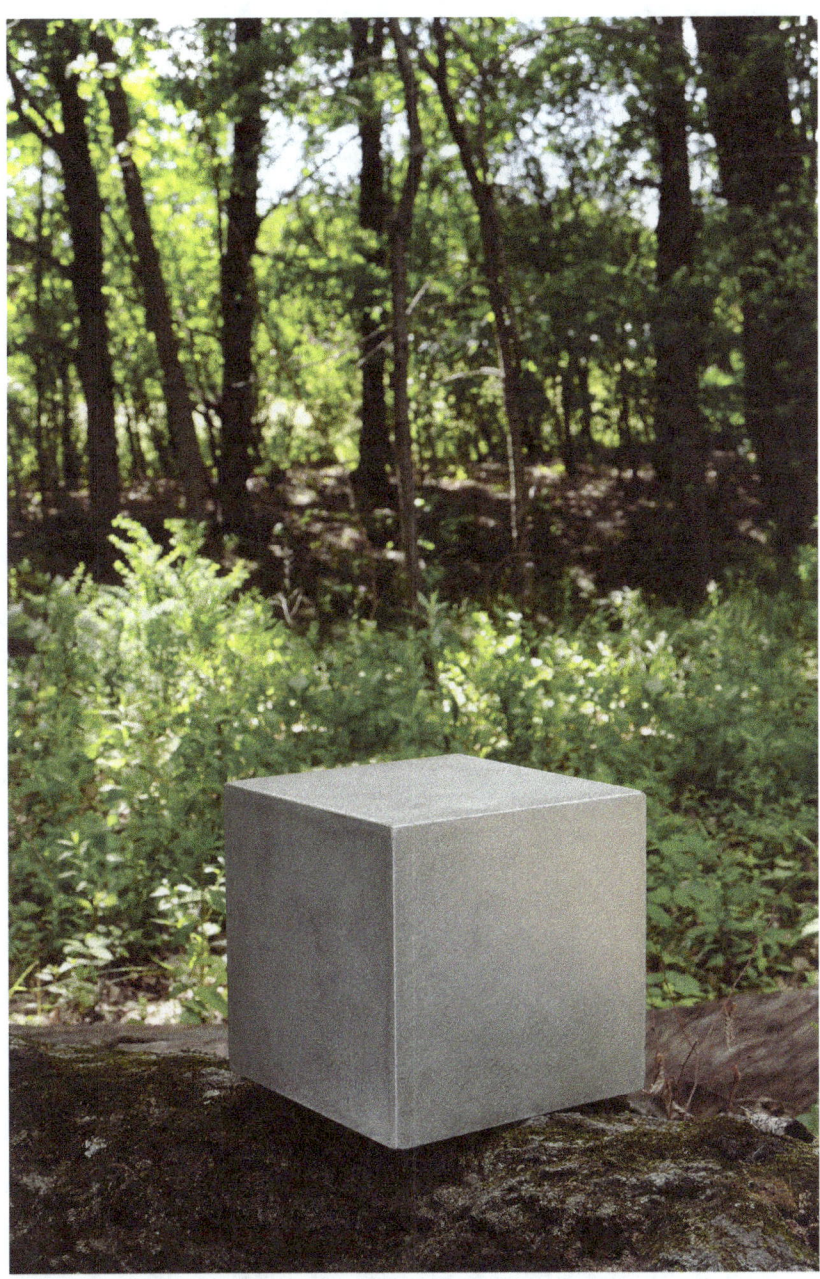

Figure 1.4. Aluminum cube, 12 × 12 × 12 in.

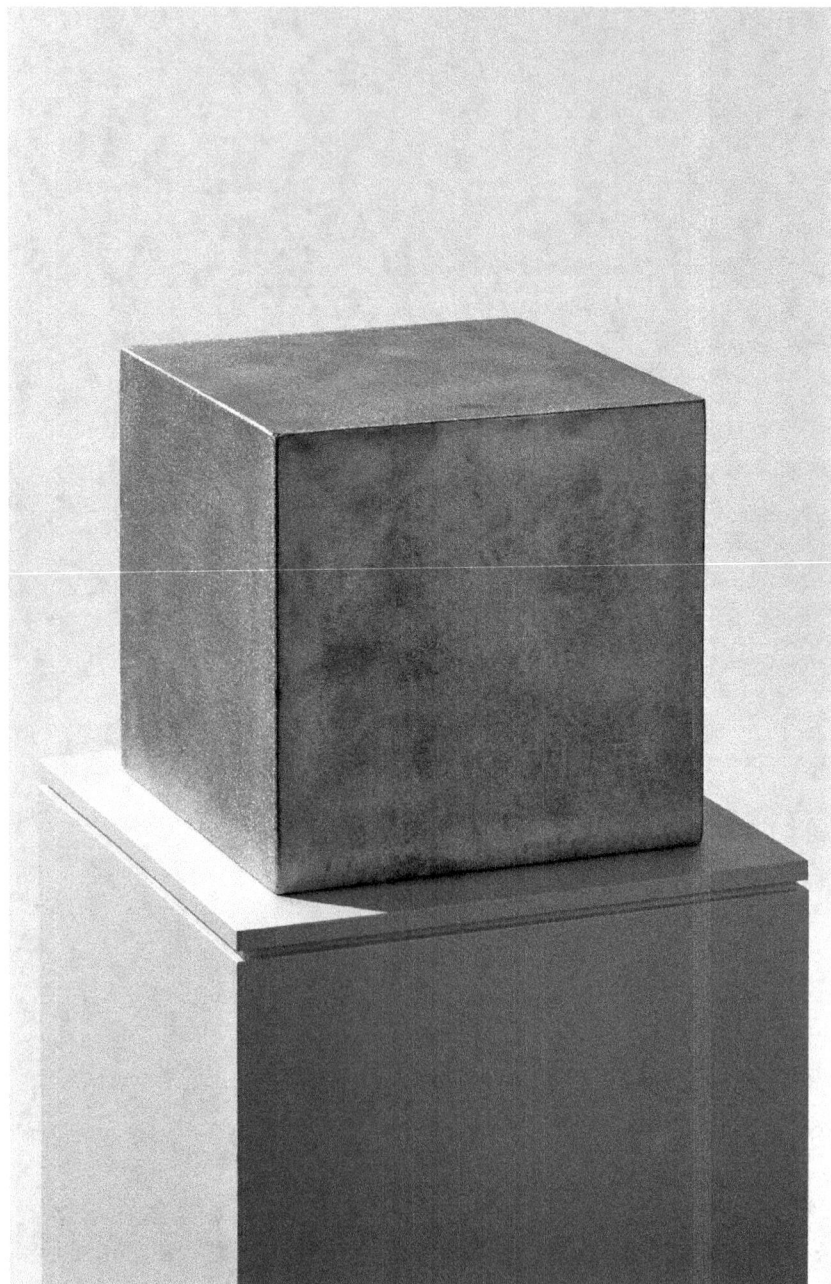

Figure 1.5. Aluminum cube, 12 × 12 × 12 in.

Awareness of place suggests a related concept, the idea of "normal" place. We come to know objects within certain contexts; that is, we expect certain objects to be in certain places. Books, for instance, belong on shelves and on desks and tables; toothbrushes, in a bathroom. Over the course of history, the normal place for works of art has been fluid, changing over time. Sculptural objects, once placed on pedestals, now stand or lie or hang free. Conventions change, indicating cultural shifts in thinking, so that an object once placed inside a building, on a pedestal, might today be found outside, in a forest perhaps. We do not see this object as being in an abnormal place; rather, the normal has grown or shifted to include that which would previously have been considered abnormal.

What Is It Made Of? How Is It Put Together?

All objects are made of some material or combination of materials formed or put together in some manner. Locked within these materials and methods is a history, a history of transformation, of change over time. Each material speaks the language of its physical characteristics—of its texture, durability, and color, for example. The processes by which objects are formed are often hidden from immediate recognition. The wood of the table hides, in a sense, the secrets of the table's creation from all but the most informed. We forget, or better, do not see, the tools and techniques of felling, transporting, milling, drying, and assemblage. Once wood has been assembled into a recognizable object, a table, it is oftentimes easier for us to see and understand the scratch on the surface, the dent on the leg, or the ring left by a wineglass, obvious reminders of past events. The future also speaks to the attentive. We can imagine a prized table, over time, protected and removed from overt danger and thus not succumbing to any dramatic physical changes. But wood may burn, and styles will certainly change. That which was once thought worthy of special care and attention might, at some future time, come to seem of little value and be discarded. In a way, the past, often hidden from active consideration, is like the future—obscure. But whereas the past can usually be understood through attentiveness or research, the future can only be guessed at. Some guesses—assumptions, really—can be made with assurance or at least a high degree of certainty: iron will surely rust, glass may break, cloth will most likely disintegrate. Each object, at this moment, brings its past ahead to us in time (fig. 1.6) and activates the future in some manner (fig. 1.7).

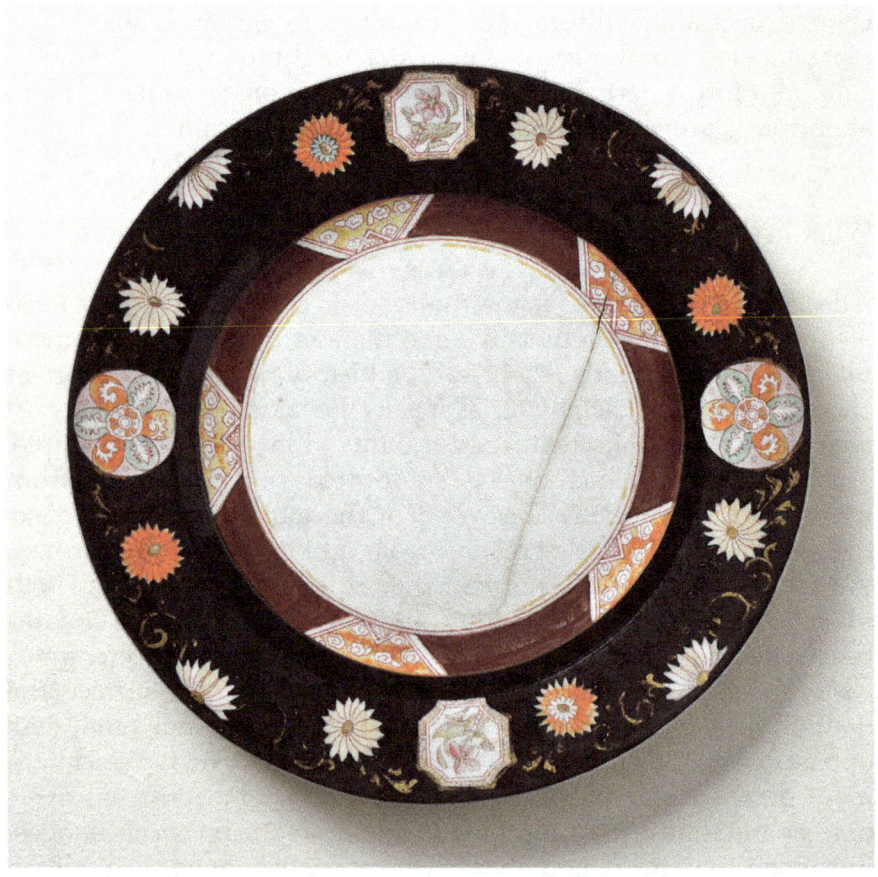

Figure 1.6. Ashworth Brothers dinner plate, 1850s–1860s. Hand-painted ceramic with traces of original gilding and prominent hairline crack, 1 × 9¼ in.

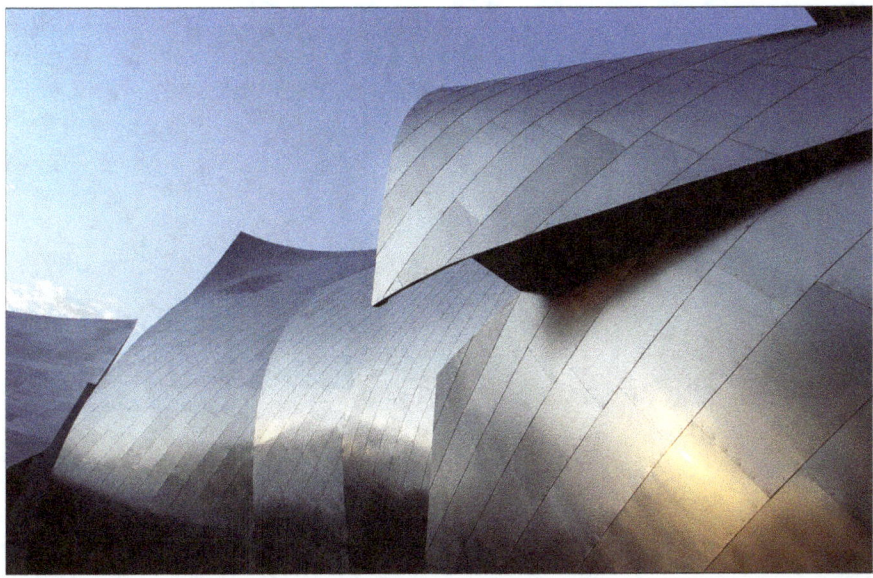

Figure 1.7. Titanium-clad roof, Richard B. Fisher Center for the Performing Arts, Bard College, Annandale-on-Hudson, New York. Photograph by Andy Wainwright, 2011.

Is It Stable?

Every object, regardless of size, material, or location, relates to gravity in some fashion. Our relationship to gravity started at our birth. Held by our parents, we were relieved of the necessity of supporting ourselves, but over time, with our first steps, we experienced the physical nature of this force. As we grew, we developed quite naturally a sense of equilibrium and a desire for stability. The force of gravity continues to teach us lessons throughout our lives, as we fall or nearly fall, reach for a heavy book, or drop a glass that then shatters. We bring such experiences and lessons, learned and relearned every day, to our understanding of objects.

Let us now look at an object. It measures thirty-six inches in length and is made of brass, a nonrusting metal commonly used in industry. Let us now place this object in three different orientations—horizontal, vertical, and diagonal—and see how changes in position affect our understanding of the object.

The question is, Is it stable? In the first image, with the bar lying horizontally on the floor (fig. 1.8), we can see that the object is very stable. Gravity is pulling the bar down evenly, and there is very little chance that it will be physically upset.

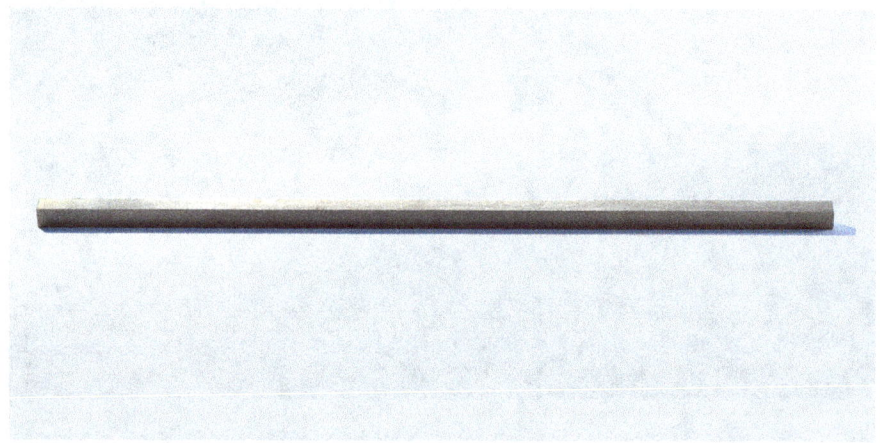

Figure 1.8. Brass bar, 36 × 1 × 1 in.

Looking now at the next image, we see the bar standing vertically (fig. 1.9). Let us now ask the same question: Is it stable? In this vertical position, with such a small base to stand on, the object could easily be upset if we bumped into it. It stands alert but is ever in danger of falling. It is stable, but for how long? you might wonder.

If you were asked to place the bar in a diagonal position, you might accomplish the task by leaning the bar against a wall (fig. 1.10). You would realize that without the support of the wall, the bar would surely fall. The bar cannot physically stand between the vertical and the horizontal without some such aid. With this aid the bar is quite stable, but without it, gravity would immediately cause the bar to fall.

Every object is more stable or less stable. When experiencing objects, we intuitively feel their degree of stability and make appropriate emotive associations. For instance, a very stable horizontal object causes us to feel the calm inherent in this orientation, while

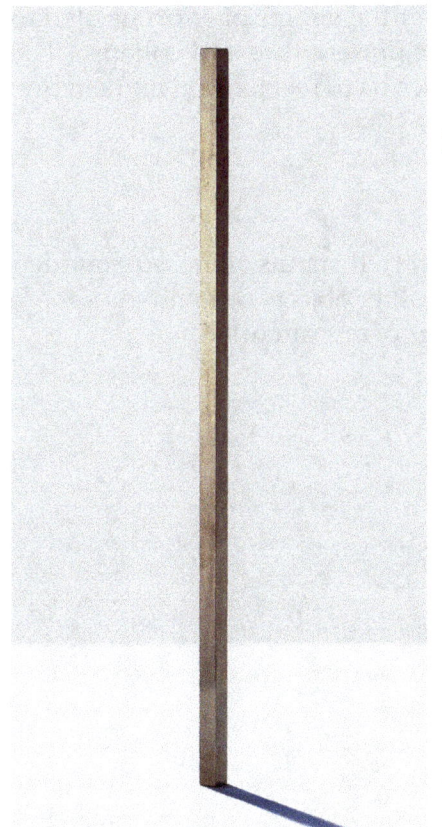

Figure 1.9. Brass bar, 36 × 1 × 1 in.

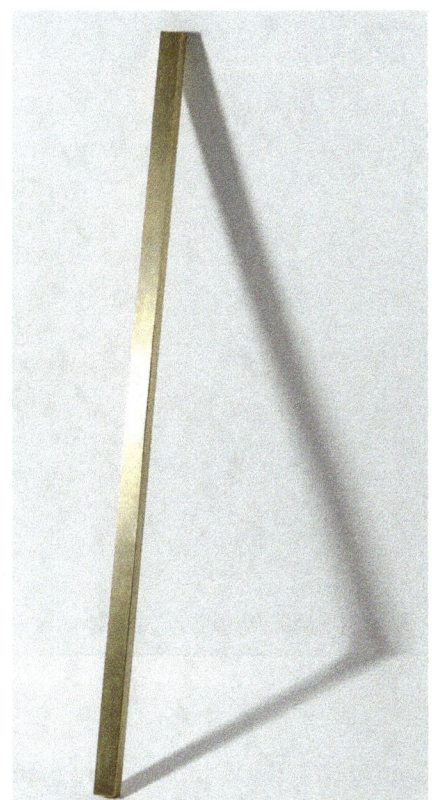

Figure 1.10. Brass bar, 36 × 1 × 1 in.

a vertical object causes us to consider the degree of effort needed to maintain this position and the ever-present threat of collapse. The diagonal's inherent instability causes us to feel tension and energy, perhaps excitement.

How Many Are There?

Let us look at a brass sphere (fig. 1.11). It stands alone, surrounded by empty space. Since there are no other objects present, the sphere is the obvious focal point, the center of our attention.

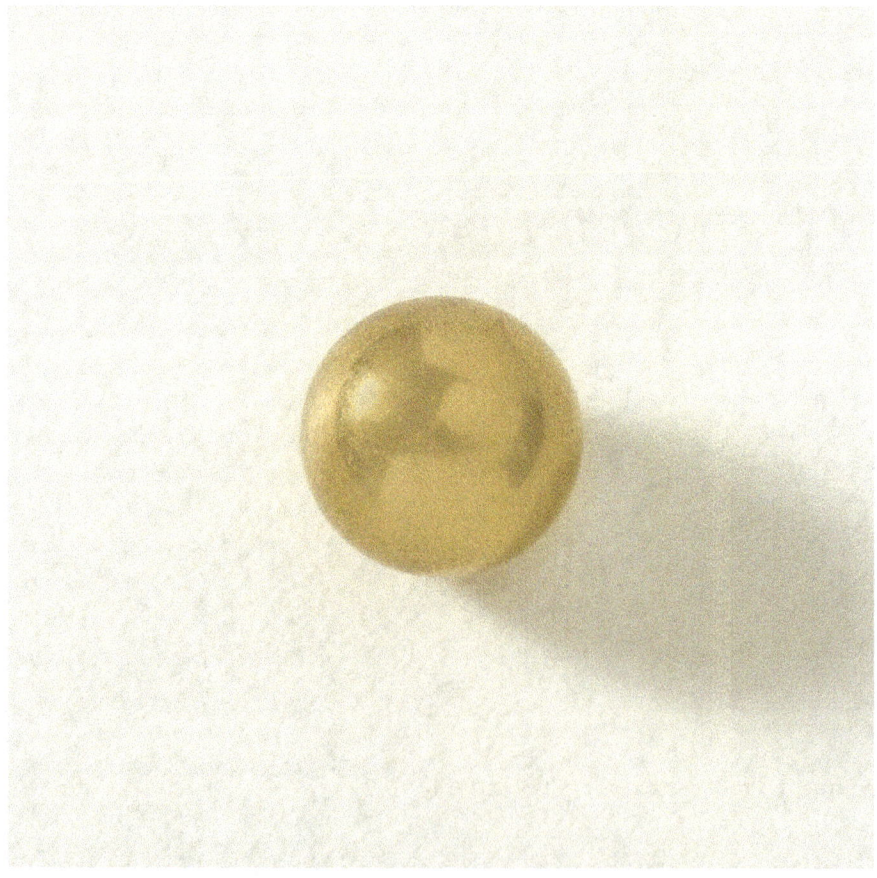

Figure 1.11. Brass sphere, 1 in. diameter.

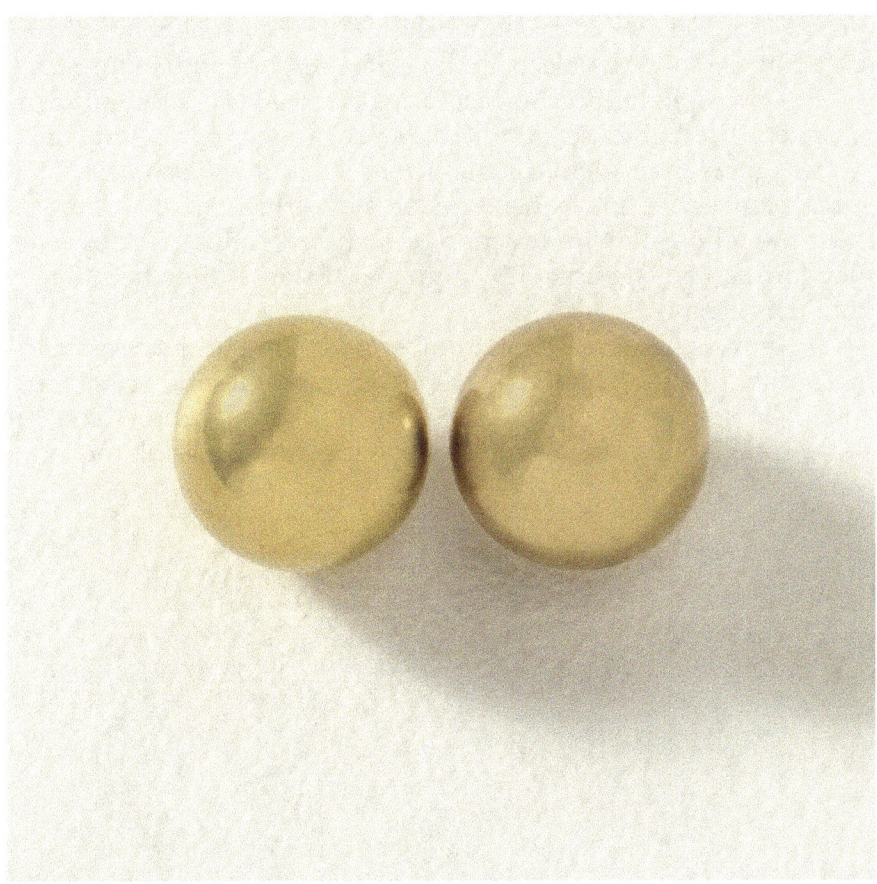

Figure 1.12. Brass spheres, 1 in. diameter.

Let us now look at another image, this one presenting two brass spheres (fig. 1.12). They rest side by side, close to each other. Like the solitary sphere in figure 1.11, these spheres also create a single focal point. Though each is a separate entity, their proximity and shared physical characteristics unite them, drawing them together somehow.

Let us return to the solitary sphere and consider it. The sphere stands alone in that empty space—no doubt, no ambiguity, just presence and definiteness. Think about other solitary presences—a church on a hill, a monument to the lost, a book held in your hands. How definite they all are. Each demands attention and is seen as we often see ourselves, as an individual. With no visual competition, each of these solitary objects declares its presence with assurance.

The two spheres, in contrast, bring us to a very different mental state. Paired, they seem united and separated at the same time. We recognize their identical characteristics, but we also notice the space between them. Depending on how much space separates them, we will feel their attraction or their separation. Side by side, they form a pair; separated by a greater distance, they become individuals again.

If we were to increase the number of spheres, our understanding of them would change. Depending on the number of spheres and their spatial arrangement, we would draw various conclusions. Imagine a large number of spheres randomly placed on a

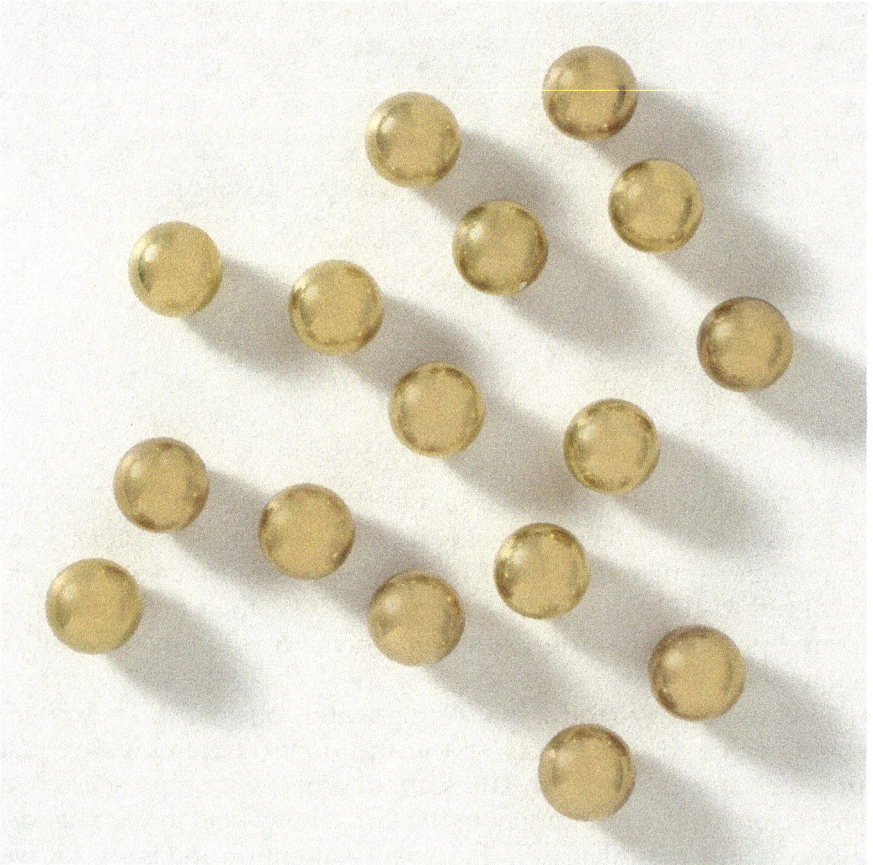

Figure 1.13. Brass spheres, 1 in. diameter.

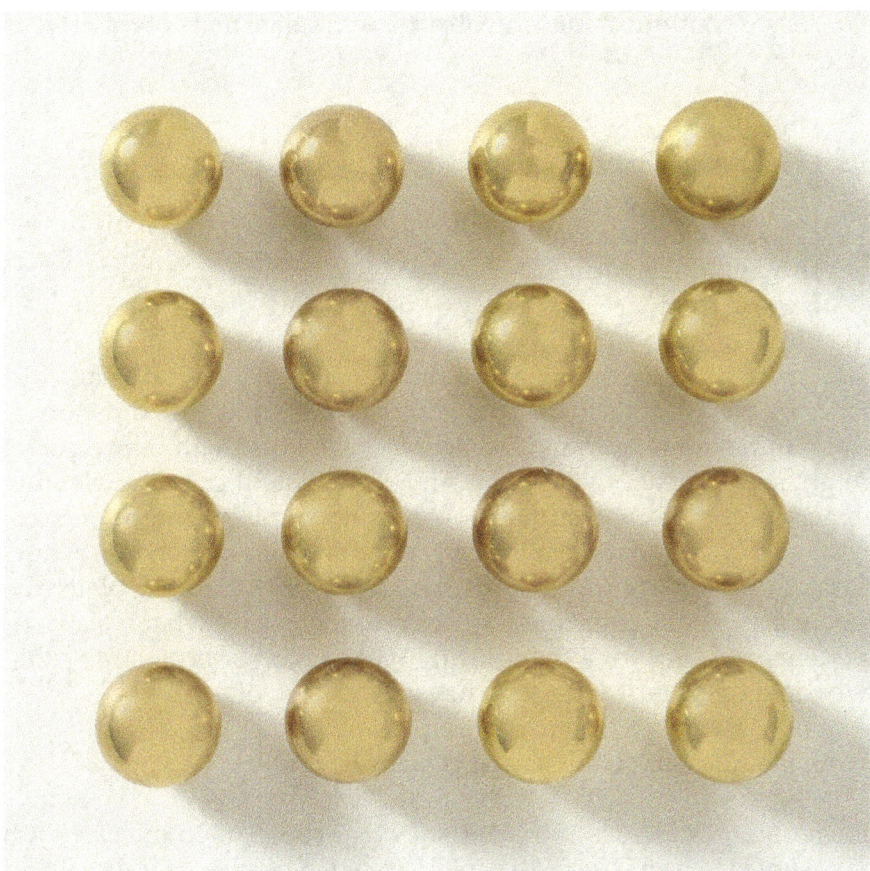

Figure 1.14. Brass spheres, 1 in. diameter.

surface (fig. 1.13); now imagine the same number placed in a grid pattern (fig. 1.14). The viewer would make obvious associations in each scenario: chaos, chance, perhaps freedom in the one; order, stability, perhaps regimentation in the other.

In thinking about number and arrangement—or, for that matter, about any other physical characteristic—we bring our understanding of our physical and cultural world, the world that we have been navigating since birth, to our understanding of works of art. Creative artists, responding to and sharing this same physical and cultural reality, intuitively follow patterns, eliciting responses that allow for our shared understanding.

As we continue exploring objects, we realize that we might also ask

How many colors?
How many shapes?
How many textures?
How many directions?
How many materials?
How many places?

We can ask the question, How many? of any physical character-istic. In responding to the questions, we must become aware of the ramifications of the artist's choices. A work with many colors might seem very active; a uniform texture might calm the viewing experience; a work occupying two different places might create a dialogue or perhaps a feeling of separateness and isolation. A suc-cessful work of art does not disclose its identity easily or blatantly. There are usually multiple ways of understanding what we are ex-periencing. It is with a thoughtful investigation of possibilities that we come to understand the richness of a work of art.

How Do You Physically Interact with It?

Each object possesses characteristics that cause us to interact with it in some fashion. We readily pick up a pen because it is small and functionally identifiable. We pass our hands over a sweater know-ing that it will be soft. We pick up broken glass carefully because we know that the shards may cut us. Throughout our lives we learn lessons through our interaction with the physical world. We touch ice and feel cold; we slip and feel our loss of balance; the sun warms our face and the rain dampens our clothes. Over and over again the smooth surface is smooth and the jagged edge remains jagged. This cup fits comfortably into our hand, while the edge of that table cuts into our arm. Certain smells attract us; others repel. These lessons, continually reinforced, are at times refined by the introduction of new information. This edge may feel slightly different from that edge; this smooth surface feels like glass, while that one feels like wood. Through cultural conventions, we learn that in a museum we are not to touch the sculpture on the pedestal nor to pass our hand over the surface of a painting. Over time, with knowledge adequately learned, we can anticipate. We know, for instance, that when driving we must slow down when approaching a curve and that the lid that needs lifting from boiling water will burn our fingers if we are not careful. We now know, before entering the museum, that we must be quiet and respectful once inside.

We can see how well we have incorporated those lessons by watching students interact with sculptural objects in the classroom. Given permission to interact freely with the objects, students know intuitively how to approach them: This hanging object is gently nudged, while that one is given a strong push. Some containers are picked up and shaken, while others are only looked at. This floor piece is circled, but that one is stepped through. We put our noses to one surface and smell it, while never dreaming of doing so to another surface. Our competencies seem staggering. Quite naturally, it seems, we interact with sculptural objects in an appropriate and similar manner. In a way, we have all learned the same lessons and have generally learned them well.

What Similarities and Differences Do You See?

Here are two objects, presented side by side, close to each other, sharing the same place, at the same time (fig. 1.15). Both objects are made of metal. Both are cubic in form. We do not know the objects' actual size, but we do know that they are the same size. We see a commonality of characteristics, but also marked differences. The one object appears to be a solid metal block, but since we are looking at a photograph, we are unable to say definitively that the block is solid. If we could experience it, we could employ other senses besides sight to understand its true nature. We could try to lift it to determine its weight; we could pass our hands over the surface to feel its temperature and texture; we could examine its edges and surface closely to ascertain the method of construction; we could tap it, and listen.

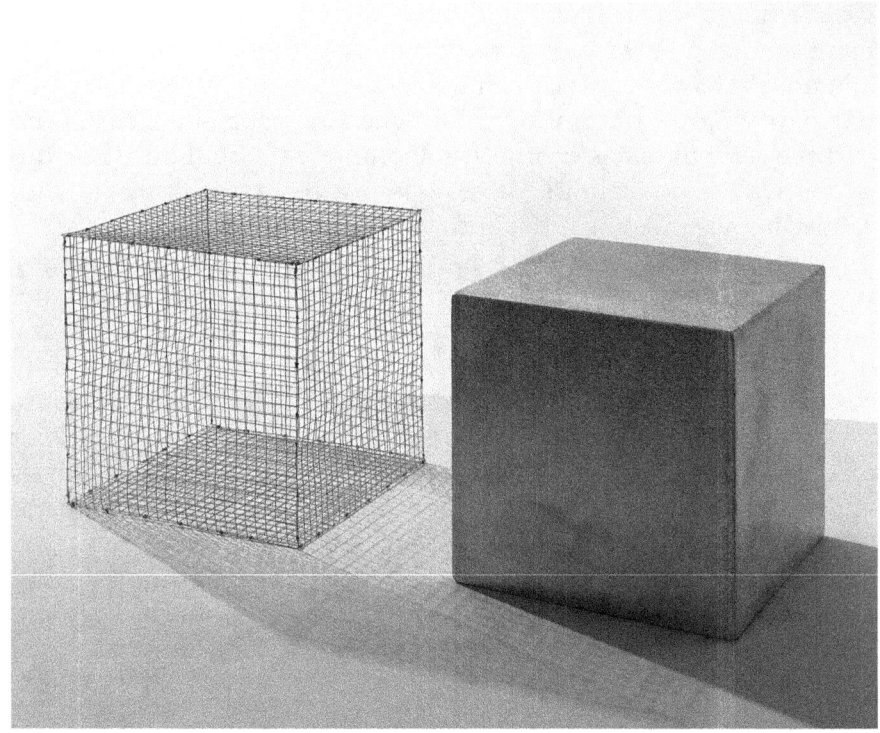

Figure 1.15. *Left,* galvanized steel hardware-cloth cube, 12 × 12 × 12 in.; *right,* aluminum cube, 12 × 12 × 12 in.

Let us assume, however, that the block is solid. With an appropriate knowledge of metals, we would recognize it as aluminum, commonly used in industry for its strength, light weight, and durability. At present, the block appears very stable, and given its material qualities, we can be reasonably assured of its longevity. If we imagine its future, say, in one hundred years, we can be relatively certain that it will exist in much the same condition as we now see it in. Its past, present, and future are, in a sense, united, sharing the same physical qualities over time.

In contrast, let us look at the second object, the cage-like structure. Again, with an appropriate knowledge of metals, we would recognize galvanized steel hardware cloth as the material used in its fabrication. This metal is also very stable and unlikely to deteriorate over long periods of time. But is this object's future equally secure? Would this object survive the next one hundred years? We might pause. Those thin walls could easily be crushed. We recognize, in fact, a quite fragile object here, one whose future is not physically secure.

We see from these examples that we make assumptions about the future when experiencing objects in the present. We may not actually be aware of making such judgments, but our minds are registering them. As our knowledge of materials and processes increases, we become more and more capable of making these kinds of judgments.

The aluminum block, in form and material, approaches an ideal state. We are led, as observers, to geometry, timelessness, and perfection. The cage-like structure, much more subject to real-world forces and possessing characteristics that remind us of the real world, causes us to think of the here and now, the temporal world of utility and change over time.

These two objects, for all their shared characteristics, are thus essentially quite different.

Some Additional Questions We Might Ask

- **What Is the Role of Color?** Whenever we encounter an object, we experience its color or combination of colors. Color is critical to our understanding of objects. Color can visually unify or break up a form; disguise or hide an inherent surface; lead us rhythmically; evoke emotions; or create physical sensations. We must remember, also, that some aspects of color appreciation are culture-specific and can be understood only through research. Our intuitive sense, therefore, must at times be called into question, especially when we encounter objects from different cultures or times.

- **Does the Object Move?** All objects assume an attitude toward movement. Some objects are immovable or very nearly so (think of a mountain or a building); others are seldom stationary (think of a feather floating in the air). Those objects that do move may move very slowly, like a lava flow, or very quickly, like rushing water. Movement, both actual and implied, has characteristics that evoke physical and emotional responses. Whether continuous and repeated, or interrupted and unique, movement is vital to an object's identity.

- **Can You Enter the Object?** Imagine how many times you have entered a building, opened a bureau drawer, or filled a glass with water. In all these cases, you were allowed to enter an object in some fashion. Some objects, however, do not allow such entry, forcing the viewer-participant to remain on the outside (think, for example, of a rock). Whether we are allowed inside an object or are forced to remain on the outside, the repercussions are significant to our experience and understanding of the object.

- **How Do You Visually Move Through the Object?** All objects present an array of visual characteristics. When we view an object, our eyes are often led from one characteristic to another, with some aspects assuming greater importance than others. Oftentimes a visual contrast in color or size, or perhaps a certain shape, causes us to proceed in a specific fashion. In a sense, we follow a hierarchy of interest and come to know the object in this dynamic way.

What Is the Object's Cultural Foundation?

In September 1921, for the exhibition *5 × 5 = 25: An Exhibition of Painting,* the Russian artist Alexander Rodchenko exhibited three painted canvases titled *Pure Red Color, Pure Yellow Color,* and *Pure Blue Color.* As their titles indicate, they were painted pure red, pure yellow, and pure blue. They were moderate in size, measuring about twenty-four inches by twenty inches, and were evenly painted. Imagine, for a moment, if we could somehow see these paintings again, experiencing them directly, as the viewers did in 1921 (fig. 1.16). What would we think? What could we logically say about the paintings?

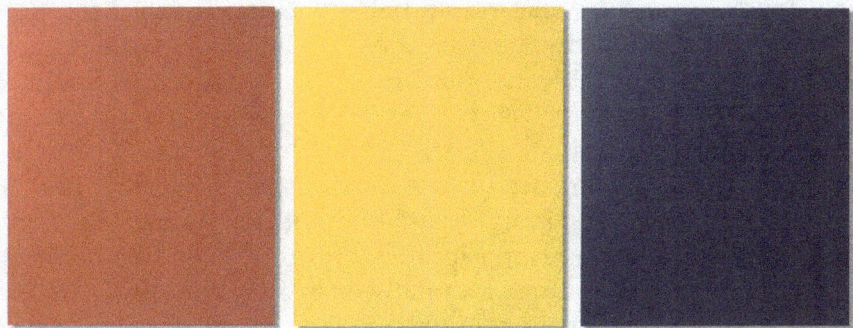

Figure 1.16. Facsimile of Alexander Rodchenko, *Pure Red Color, Pure Yellow Color,* and *Pure Blue Color,* 1921. MDF, latex paint, 24 × 20 × ½ in. (*each panel*).

Undoubtedly we would recognize the colors red, yellow, and blue—the primary colors. All other colors derived from pigments can be created, in theory, using these three hues. In this way, the primaries are unique, fundamental, and essential. These qualities would probably seem important to us if we were trying to understand our viewing experience. But beyond this, what could we say? We might appreciate the paintings for their colors alone, finding them beautiful perhaps. We might recall associations that we have when viewing each color. But we would probably ask, Why did the artist make these paintings? In themselves, simply as colors, the paintings would not seem significant to us. We would probably feel that we needed more information if we were to understand them as works of art.

And our intuition would be correct. Art is created within cultures. Works of art are cultural products with cultural significance and, as such, cannot be understood solely in physical terms. To fully understand a work of art, we must understand its cultural aspects. The physical aspects that we have been discussing—materials, methods, size, place, number, and all the rest—are in themselves inadequate to fully explain what we are experiencing.

In the case of the Rodchenko triptych, one must return to revolutionary Russia. Artists at that time were passionately debating the role that was envisioned for the artist in the new communist society. At issue was the individual artist, intuitively "composing" works of art and exhibiting them within a system of galleries, for individual edification and profit—much as in our culture today. Revolutionary artists wanted to overthrow the old artistic ways—the practices, institutions, and products of the old Russia. Intuition would be replaced by rational decision making; private galleries, by museums; individual desire and self-aggrandizement, by societal needs and responsibilities. Painting would be replaced by the constructed product, rationally conceived and designed and serving a practical, functional purpose. In time, avant-garde artists would come to make fabric patterns for dresses, invent machinery for industry, and design woodstoves for the home. It was within this context that Rodchenko painted his canvases, as a declaration, really. Painting, and all that it had entailed and stood for in prerevolutionary Russia, was finished, replaced by a new artistic purpose, dynamic, and product.

Properly seen in this context, the Rodchenko panels symbolize not only the death of painting but also the birth of the new order. Birth and death, like pure red, pure yellow, and pure blue, unique, fundamental, and essential. The seemingly insignificant has become charged with meaning and purpose. What we needed was an understanding, not only of the physical logic of the paintings, but of their cultural logic as well. Together, in dynamic interrelationship, these two logics create the significance we search for when we view a work of art and try to understand what we are experiencing. But the task can be difficult; for though the physical logic can be grasped quite easily, the cultural logic, if it is to be understood at all, must be learned through research and study.

PART II

Creative Projects

Creative Projects

The following series of projects was designed specifically for a semester-long undergraduate course in the fine arts. Though some of the projects are intended to be completed in one week, others are designed to evolve over several weeks. The teacher must be attentive to the progress and needs of each student and make adjustments to the schedule as necessary; the success of the student should be the overriding concern.

Assignment 1: Connecting Two Squares of Tar Paper

We begin the creative projects with two five-inch squares of tar paper (fig. 2.1). Tar paper is a very common building material used primarily as an underlayment, sealing out moisture on roofs and exterior walls of buildings. It is manufactured by impregnating kraft paper with tar. It is not a material generally associated with art making, and it is for this reason, in part, that it has been chosen for the first project. It is also inexpensive and easily cut, pierced, folded, or rolled.

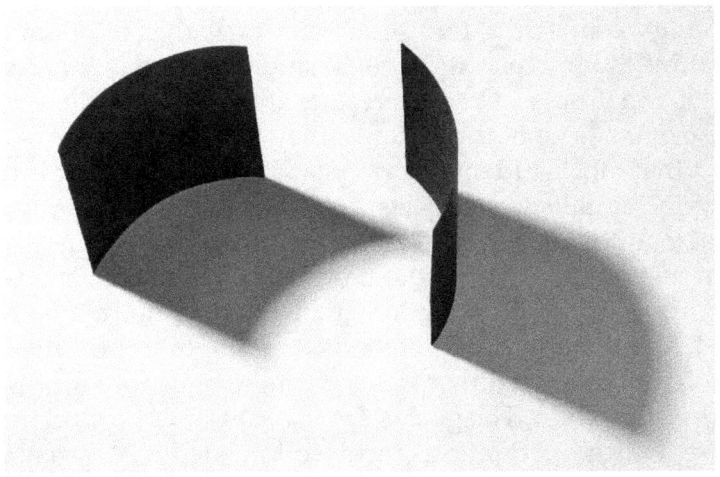

Figure 2.1. Tar paper, 5 × 5 in.

For the first assignment, students are asked to physically connect the two squares. The goal will be to connect them in a very interesting and adventurous way. Standard fashions of attaching, such as stapling and gluing, are to be avoided. We will be looking for unusual approaches, exciting directions. The squares must remain black; they cannot change in color or be decorated in any way. We want them to remain identifiable as tar paper. Students are allowed to manipulate the tar paper in any way they wish, as long as their actions are directly related to the attaching system. They cannot arbitrarily cut, pierce, or fold the squares. Any action performed must be integral to the attaching system. Flexible or rigid elements may be used to join the two pieces. Students should create a fully three-dimensional presence. Objects should not be attached flat to the wall, for this would be considered a two-dimensional approach. The created object may be placed anywhere; it need not sit on a table. The final product must be appropriately crafted.

As we begin the creative projects, it is very important to remember that students are at the start of a process that for most will seem strange and perhaps bewildering. Most have been ill prepared for the inventive activities they are undertaking. Coming from an environment of consumption and standardization, they have seldom been asked to find inventive solutions to challenges. It is important to remember this as we guide students through these assignments.

Though connecting two squares of tar paper offers unlimited opportunities for exploration, most students display inherent prejudices in the process. Creative results are often strongly influenced by the original size of the squares: they are hand-held. Also, given the fact that most students work on a table or a bench, it is not surprising to see finished projects that are small in scale and placed on a table for display (fig. 2.2). The first project is designed, in part, to demonstrate to students these prejudices. The project is also designed to promote those actions that reflect a more adventurous approach to materials and methods. It is important for the teacher to highlight accomplishments that demonstrate more sophisticated thinking and action. The goal is to move students beyond their first impulse, which often reflects received knowledge. An easy answer is usually not a good answer. Highlighting student achievements

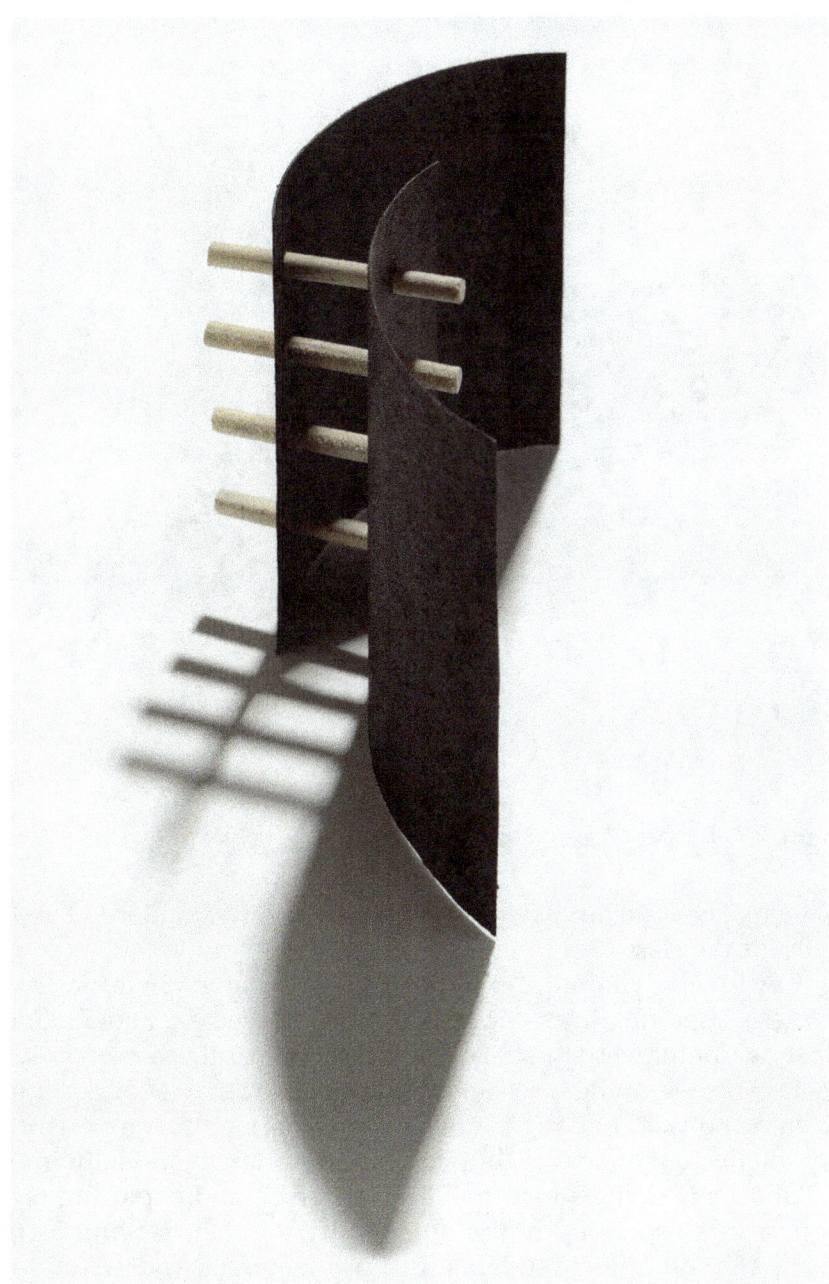

Figure 2.2. Tar paper, dowels, 5 × 7 × 3 in.

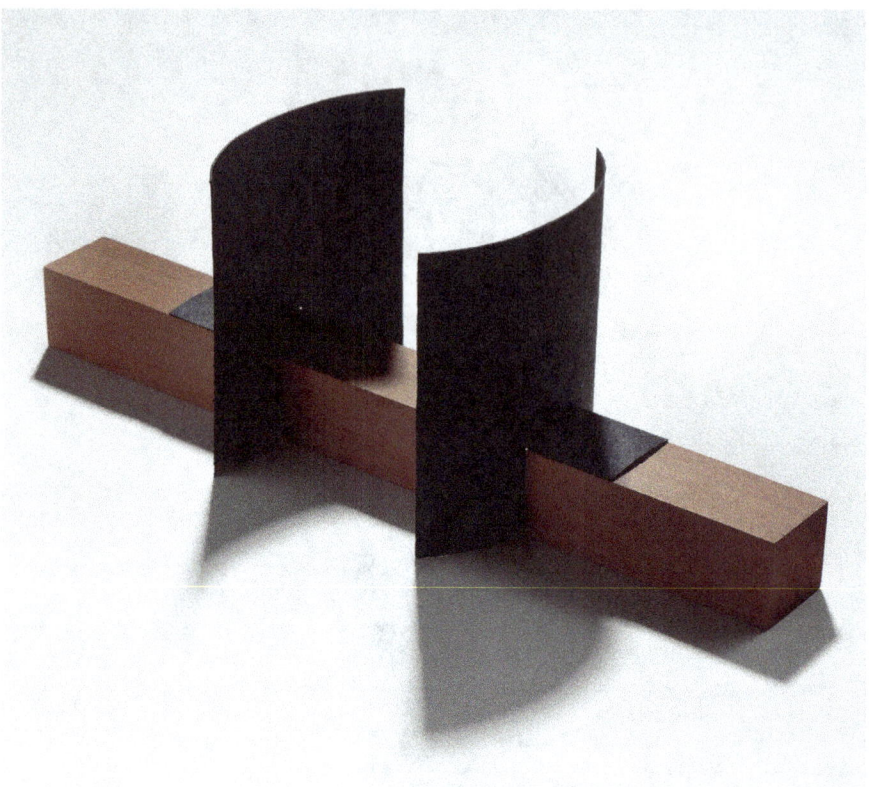

Figure 2.3. Tar paper, wood, 5 × 14 × 5 in.

that move beyond an easy response helps to set a standard of excellence for the class (fig. 2.3).

Key to this dynamic of exploration is creating an environment of richness—one rich in materials, methods, discourse, and thought. Most beginning students face the challenges of these exercises, at least in the beginning, with an intellectual bias: they want to "think up" a response. Thinking is obviously an important component of any intellectual endeavor, but equally important, especially in the visual arts, is doing—manipulating materials and exploring techniques. It is especially important that the teacher, serving as the creative guide, direct students to those materials and techniques that lie outside their common experience and range of comfort, as well as to those not usually associated with art. Under the guidance of the teacher, students must experience the classroom as an arena of exploration.

Assignment 2: Connecting Two Blocks of Wood

The second assignment is designed to overcome the prejudices often encountered in the first—those of place and scale. With the second project, we again ask students to make a connection—this time, to connect two wood blocks measuring six inches by four inches by two inches (fig. 2.4). Physically, the blocks differ from tar paper in certain important ways: the blocks can stand quite easily on their own in a variety of positions, and they have greater mass, more material stuff, to work with. Wood also presents different challenges in terms of tools and techniques.

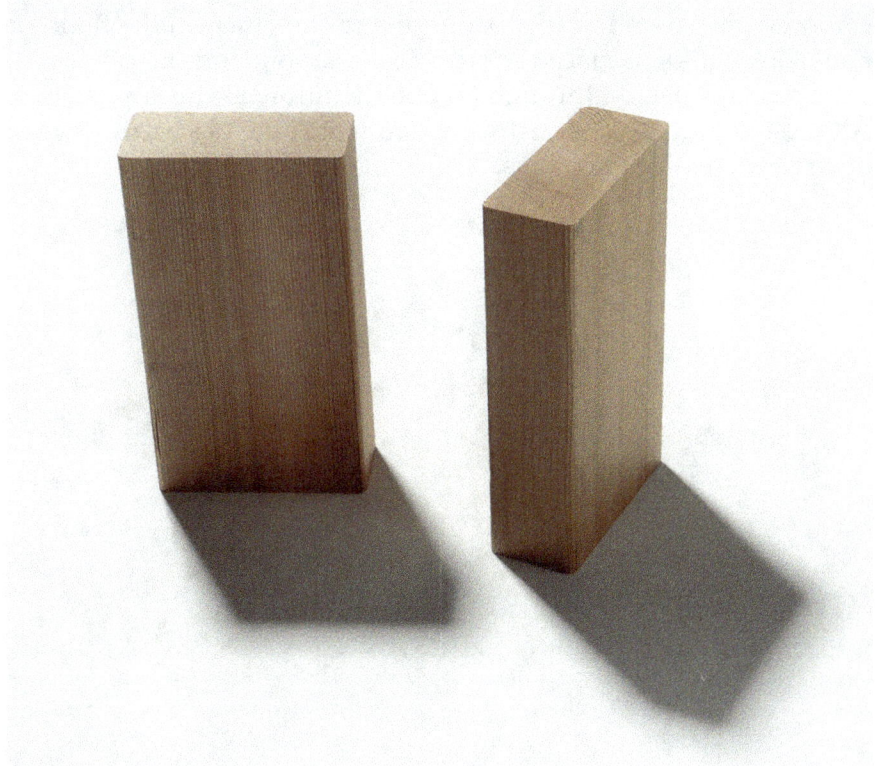

Figure 2.4. Wood blocks, 6 × 4 × 2 in.

* * *

At this point, the goal is to coax students to move outside what seems normal to them. Generally, it is quite easy to demonstrate to students that most of them responded to this assignment much as they did to the first: most projects were presented on a table and were relatively small. Students now need to be shown an alternative approach. To accomplish this, the teacher can introduce place as a generating factor in the creative process. Since most students never actively thought about place in this way, a demonstration is useful at this point. For example, the teacher can take the blocks from the table and place them on the floor several feet apart, then ask the students how they might join them (fig. 2.5). The distance between the blocks could then be altered to show how changes in place can bring about changes in scale. The teacher could continue the demonstration with further examples: for instance, one block could be placed on top of a pedestal and the other on the floor, or one block could be placed inside the classroom and the other outside, perhaps passed through an open window to the lawn below. All these examples demonstrate to students how place can serve to open new creative pathways.

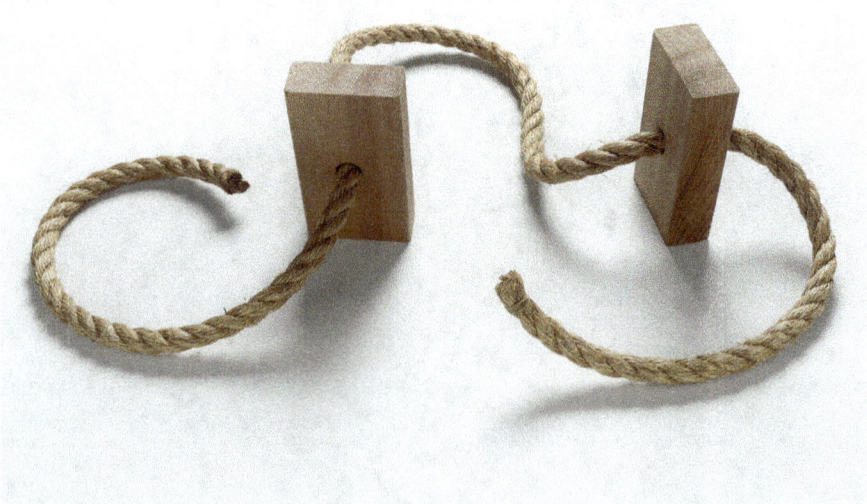

Figure 2.5. Wood, rope; dimensions variable.

Though most students find this method understandable, many find it unsettling and are often reluctant to use it: it's not the way they are accustomed to working. Contextualizing the challenge, however, causes students to realize the importance of place and allows place to do some of the creative work for the students. Approaching the assignment in this fashion is to some degree easier: creating in a spatial vacuum locates the challenge solely in the materials and their manipulation, while accepting place as a generator offers another avenue of exploration and enriches creative options.

Whether place is to be a generative element in the creative process or not, it is important for the teacher to encourage the use of a wide range of materials, as in the first assignment. An excellent way to accomplish this is to present students with a material and then ask them whether they could imagine a connection using that particular material. Such a demonstration once again suggests to students the use of techniques and materials they might not normally have thought of. The success of this method is reflected in the range of interesting materials that students have used in past projects, including water, laser beams, sound, and a living creature—a dog, who wore the two blocks in saddlebag fashion. A class exposed to such a wide range of materials is much more likely to develop an adventurous approach to problem solving as the course progresses.

Assignment 3: Altering a Plaster Block

We now turn our attention to an object measuring twelve inches in height and made of plaster, a very hard and brittle material (fig. 2.6). The object is quite heavy—if you held it in your hands, you would feel its weight. As presented, it stands upright and, with such a substantial base to rest on, is quite stable. It is presented alone, with no other objects around it, in a neutral space, a space with few identifying characteristics. The object—a rectangular solid, a geometric form of straight lines and uniform right angles—is uniformly white, its inherent color. It appears pristine, pure, without blemish, with no evidence of its creation or history. It is here, in the present. For our purposes, let us now assume that we are with this object.

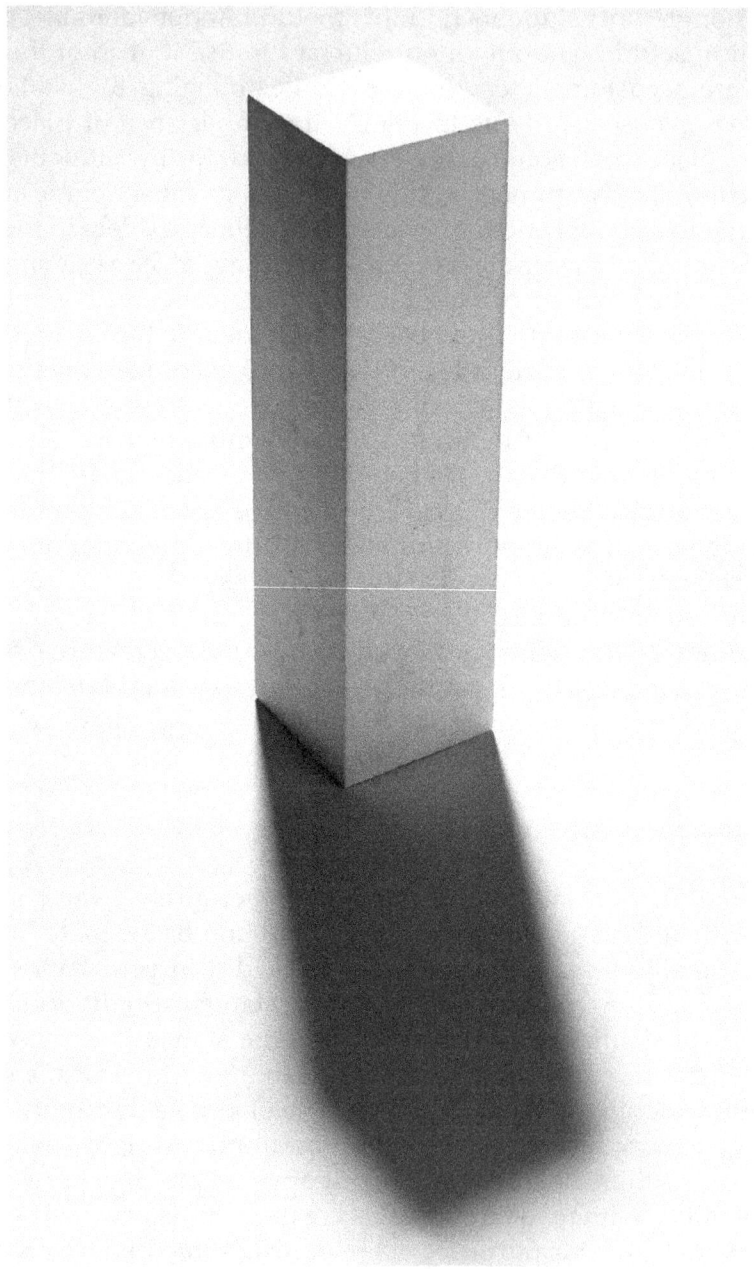

Figure 2.6. Plaster block, 12 × 3 × 3 in.

For the next assignment, each student is given a plaster block and asked to change it in such a way that its essential nature is radically altered—changed not slightly but fundamentally. Students may add what they like or subtract as they wish. They cannot alter the color; the block must remain identifiably white. The final product must remain visible to the viewer.

How could students logically approach this creative challenge? Let us return to our questions, those same questions that we used to structure our interpretive investigation of objects, and now use them as creative tools capable of opening up possible avenues of exploration and invention.

Let us think about number: How many? We now have one, but through cutting or breaking we could have two or four or, for that matter, a hundred pieces. The block is very dense, but if we were to break and then grind it to a fine powder, not only would its density change, but its weight, or at least its perceived weight, would also change. Those cut pieces or fragments or dust-like particles could stay tightly bunched together or could be separated and spread out, thereby changing not only the block's size but its location as well. What was only here could now be here and also there. The block, unencumbered with story beyond its physical aspects, could become an actor in a narrative. Exchanged between two people, it could become a gift; thrown through a window, it could become evidence of guilt; placed on an altar, perhaps a sacred object.

We can see, through these examples, that the questions that have unlocked visual information and significance can also be used to unlock creative options. Our method of questioning can propel creative research as well. Each category of questioning can be used for this purpose. The above creative examples used number, density, weight, place, and narrative. We could still approach the creative challenge through movement, containment, and dialogue, for example, quite easily. Beyond this particular example, any artistic challenge can be approached in this fashion. Instead of waiting for an intuitive response, the creative artist can direct research very rationally and systematically, using the physical categories that we discussed in the first section of this book as avenues of exploration.

The following is offered as a creative guide—a method of exploration.

- If there is one, make many.
- If it is here, put it there.
- If it is stable, upset it.
- If it is small, enlarge it.
- If it is straight, curve it.
- If it is whole, fragment it.
- If it is stationary, cause it to move.
- If it is this material, make it that material.
- If it is closed, open it.
- If it is free, contain it.
- If there is order, create chaos.

Approaching creative work in this logical fashion would seem to deny the value of an intuitive response. Intuition, as a form of knowledge, seems independent of rational thought, a form of felt knowledge, mysterious, almost revelatory, offering insight and direction to the inventive mind. Creative artists often rely on their intuitive powers to guide their search for new directions and forms. We might imagine that the insights gained are independent of learning, mysteriously bestowed, as gifts, by nature. We could argue, however, that these insights do have sources in the learned environment, but that these sources remain largely hidden. Our childhood experiences, for example, may play a role in forming the foundation for later intuitive decision making. A child's mind is particularly sensitive and malleable as it experiences a complex physical and cultural environment. Generally, led and directed by adults, the child willingly and unquestioningly assimilates the beliefs, attitudes, and reactions of its family and community. As childhood experiences give way to those of adulthood, patterns of behavior are reinforced. Some seem right, others wrong. For the creative mind, this sense of rightness and wrongness can become a liability, a handicap to be overcome. That which seems intuitively right could in fact be a patterned, familiar response. But the true creative artist, one who invents new visual reality, looks ahead to the future from this foundation of the past, employing a philosophy of exploration and discovery. Logically moving through alternatives allows the rational mind some measure of control over suspect emotive responses. In the classroom, such an ethos is critical to success. Intuitive responses, logically guided, must also be augmented through the use of unexpected materials and processes and the introduction of situations and challenges that seem unusual to students. If inventive decisions are based on what seems right, that sense of rightness must expand. The foundation on which decisions are made must grow; a new sense of normal must develop, and it must come to seem "right" to students.

A Demonstration: The Paper Coffee Cup

We continue our creative quest with an unlikely object, a very common object—the paper coffee cup (fig. 2.7). We might not associate this object with the fine arts, since it is so ubiquitous, so readily identifiable. And if the goal is the invention of new visual reality, such an object might seem to make that task even more difficult, since we must, in a sense, overcome the cup if we are to create something new. The question is, how to proceed.

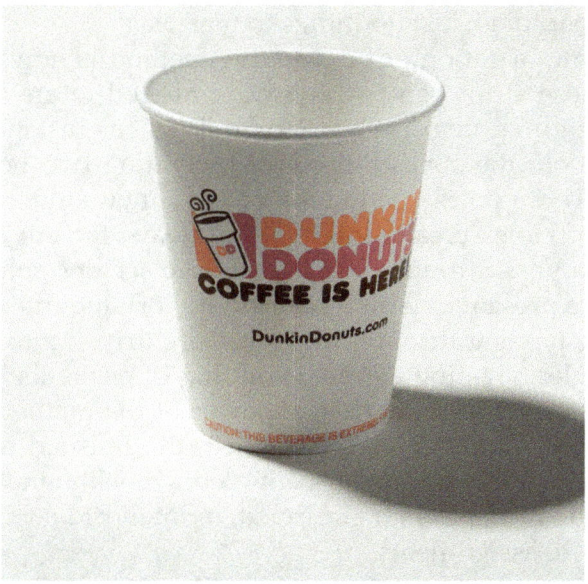

Figure 2.7. Dunkin' Donuts coffee cup.

The process begins with manipulation—the physical manipulation of the object. There are obviously countless ways to accomplish this, but for the purposes of this demonstration, the cup is cut into curved strips of paper. Using more than one cup produces any number of strips. Figure 2.8 shows these strips now arranged in a loose grouping. At this stage in the process, it is obvious to students that the known object—the cup—no longer exists, that we have overcome the cup through our actions. This new entity defies recognition as a named object, and students must now approach it as they would any new object, that is, through the physical information supplied—material, shape, number, organization, and so forth. Seeing such a transformation helps direct students away from the conceptual bias to which many of them are prone in making art. This transformation was accomplished primarily through doing, not thinking. The result was arrived at through an action, not through an idea or a concept. We are reminded of the Latin root of the verb *to invent: invenire,* to come upon. A lesson like this can have a profound effect on beginning students, for it releases them from a daunting task, one that most do not feel capable of accomplishing—the task of inventing a new sculptural presence. When students move to their next assignment, their responsibility will be radically simplified: just do things to materials.

This demonstration of the creative method is now further extended. Those strips can be physically altered in any number of ways, through cutting, bending, or rolling, for instance. For our demonstration, the strips are pierced, creating two holes, one at each end. These holes create new options, new opportunities for exploration. Holes create potential pathways for any number of materials to pass through. In figure 2.9 we see one solution. Here students are presented with a sculptural work mounted on a wall. Metal rods, passing through the holes, are driven into the wall at various angles, creating a loose grouping of elements. This single solution demonstrates clearly the power of the method. Rationally guided but intuitively accomplished, the finished work is one among innumerable possibilities. Students quickly understand the method and are thus better equipped, in intellectual terms, to proceed to future assignments.

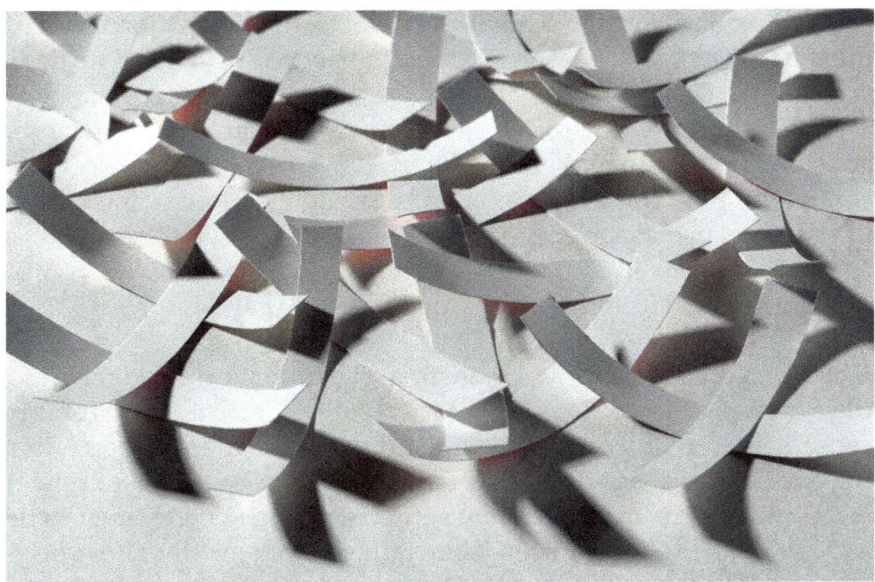

Figure 2.8. Strips cut from Dunkin' Donuts coffee cup, 5 × ½ in.

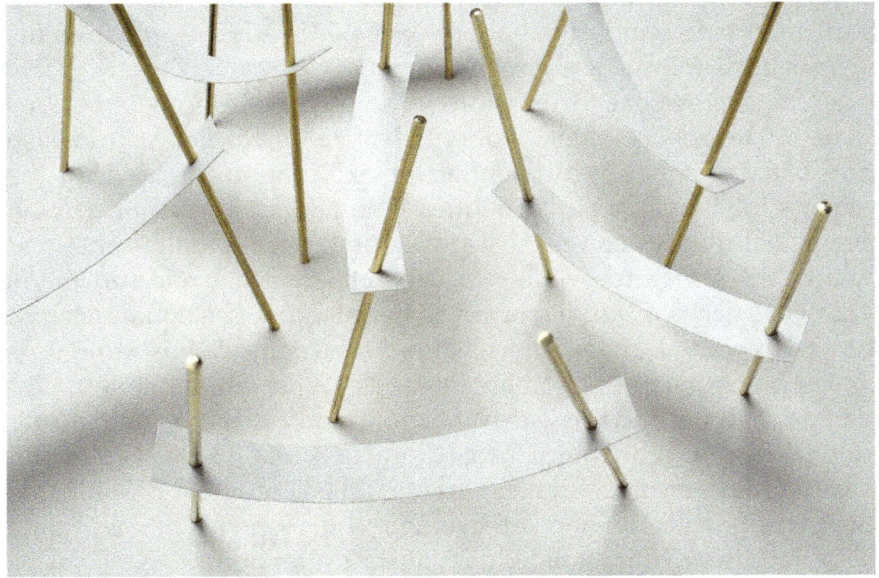

Figure 2.9. Wall-mounted strips, brass rods; dimensions variable.

Asking students to create only nonobjective works forces them to think about the various physical attributes of objects in their pure form. The task of creating these very impractical, nonobjective objects, removed from easy recognition, conceptual purpose, and direct message making, helps to coax students toward greater subtlety of thought and action and ultimately a more refined artistic and intellectual sensibility.

Assignment 4: Finding and Manipulating Paper Samples

Following this classroom demonstration, students are asked to find their own paper products. Paper is an excellent material to use because it is so readily available and comes in so many forms—like the cup. It is also easy to manipulate. Students are asked to return with at least three very different samples and are encouraged to be adventurous in their choices. They are told that for their next project they will be using one of the three paper products they choose.

On returning to the classroom, students encounter a wide range of products—paper towels, packing materials, tubes, envelopes, wax paper, specialty papers, sandpaper, papier-mâché. There seems to be no end to the variety of paper products, and students always appear impressed by the display.

What follows is a discussion of the physical aspects of various papers. The goal is to help students realize that locked within each product is potential—a potential for exploration founded on the paper's inherent physical qualities. The paper cup demonstration showed students that the curved elements were actually contained within the structure of the cup; the act of cutting released them. Likewise, each of the displayed materials possesses qualities that can be exploited in a similar fashion. Paper towels absorb liquids and can be rolled out to create linear elements of great length. Envelopes can be filled or easily taken apart. Tubes can be cut into rings. We will create objects based on such acts. These objects, unlike their known and named sources, will be new creations. Ideally, they will be visually and intellectually interesting. They will defy easy definition. They will hold a viewer's attention.

Students are told to choose one of their samples and to physically alter it. They are to create an invented object. They are also told that this object must be located in some place—a definite place of their choice. That place could be the floor, the wall, or the ceiling, but it could also be the human body or a forest. As with past assignments, students are told that place can actually come first, as a generator of options: a pond could support a floating structure; a depression in a forest could be filled; a classroom corner could support bridging elements. Whatever students' creative approach, their final nonobjective work must be appropriately crafted and interesting to look at.

Generally, students respond well to such a challenge. Of course, there are always a range of responses in terms of quality, and the teacher must continue to point out to the class those examples that demonstrate a more adventurous approach and are more intellectually challenging and interesting. And students must be shown why this is so. For example, interest may have been created through an unusual orientation or in the use of an unexpected material. Students must come to understand the source of the interest in very concrete ways if they are to develop and mature artistically. To be of pedagogical value, the teacher's judgments must always be substantiated.

Assignments 5 and 6: Exploring Paper Further

At this point, it is beneficial to draw out the previous assignment for another generation or two, asking students to further extend their research. Students are now much more aware of the physical aspects of objects and understand the power of those attributes to affect the viewer. This understanding now allows the teacher, as mentor, to guide students by suggesting creative options based on physical characteristics. For instance, a small rectangular work placed at eye level on a wall could quite logically be extended along the wall, creating a long, horizontal work. The piece could also take other orientations, vertical or diagonal. Perhaps the work could continue around a corner, dramatically altering the viewing experience. Whatever decision is made, the dynamic of invention lies with altering physical characteristics. The teacher offers various options, but the student decides how to proceed. What is most important is that the student begin to interiorize a creative method that is understandable and allows the student a large measure of intellectual control. The new piece arrives, generated through a logical sequence of decisions. Since students are at the beginning of this process, however, the teacher must not expect a full and immediate assimilation of the method. More practice is needed. To this end, students are presented with a new assignment designed to reinforce the method while limiting its scope.

Assignment 7: The Triangle Box Project

Students are now asked to bring other types of materials to class. Asking each student to bring in three very different samples again provides the class with a wide range of products to work with. As with paper, there follows a discussion that locates some of the inherent characteristics of the materials as well as various options for manipulating them. Once again, students are presented with a challenge—to create a three-dimensional, nonobjective work using manipulated materials. But now students will work from and relate to a context specified by the teacher. Shown in figure 2.10 is that unique context. It is a triangular wood box, two inches deep, with sides measuring twelve inches. Students are told to relate to this structure using their manipulated materials. The boxes will be wall-mounted and presented at eye level. Students are to create a visually interesting piece, relying solely on the manipulated materials and this structure. The box, serving both as a challenge and as an inventive instrument, will thus play a role in determining the artistic product.

Most students, at this stage, benefit from such contextual definiteness. The box helps to focus their attention and to clarify the challenge, and students generally respond with enthusiasm to the assignment.

As a preliminary step, students build a cardboard model of the box, an exact replica, which permits and encourages experimentation. Only when students are sure of their direction, usually after two or three creative generations, do they move to the wood box.

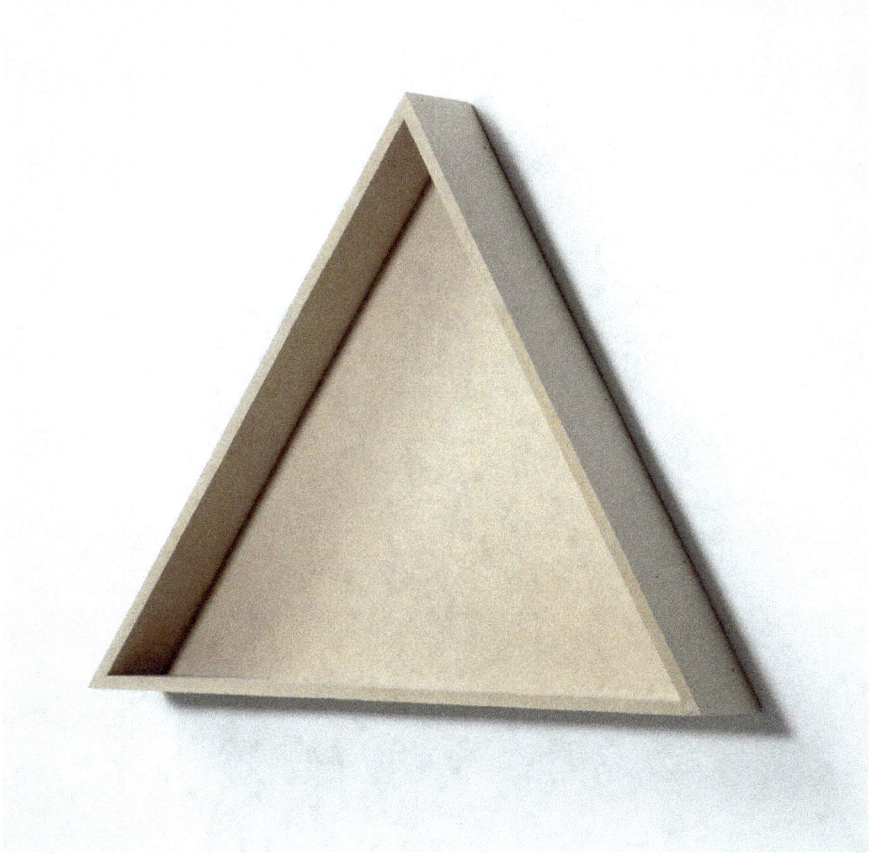

Figure 2.10. Triangular box, MDF, 12 × 12 × 2 in.

Let us now look at several examples of student work. In figure 2.11, we see a work created with the screening material commonly used in doors and windows. The student cut the material into squares and then folded them onto themselves, creating small cubic units. These units were then pierced and strung together along a flexible metal wire. After fabricating several of these strands, the student then placed them in the box, bending them to the desired form. Lastly, paint was added to the surface of the material.

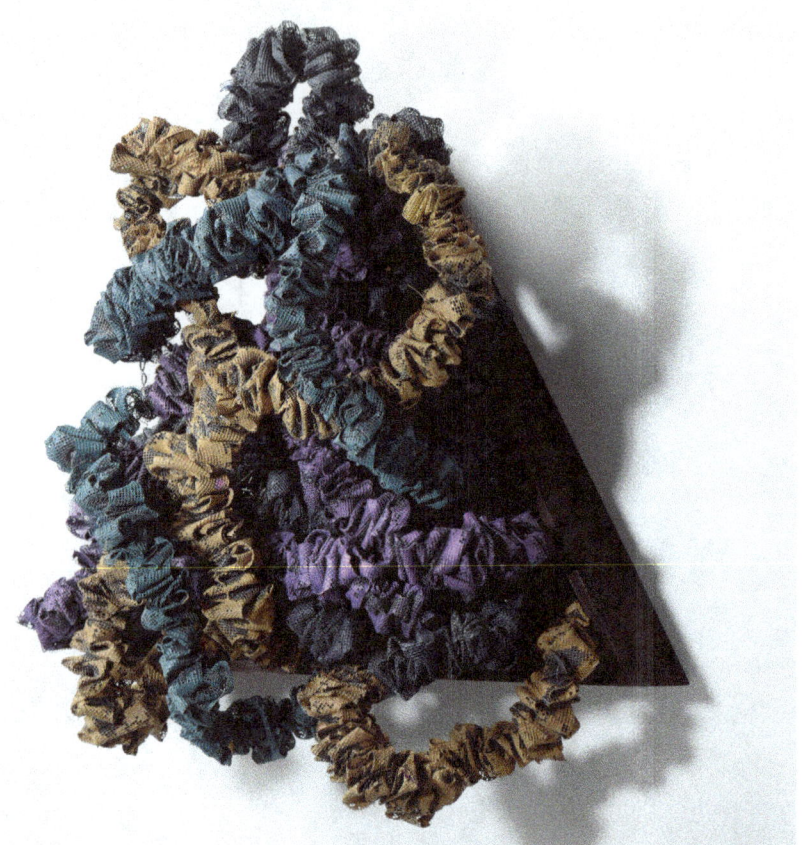

Figure 2.11. Student work, 2010. Screening, wire, paint, MDF, 16 × 16 × 6 in.

For the next work, shown in figure 2.12, the student first cut Plexiglas into various rectangular shapes, then formed them, using a heat gun, into a configuration of interlocking curvilinear units. After attaching a mirror, cut to size, to the bottom of the box, the student placed the Plexiglas assembly in the box and sealed it with a thick layer of epoxy. The presence of the mirror beneath the Plexiglas and epoxy caused a reflective glow to emanate from the work.

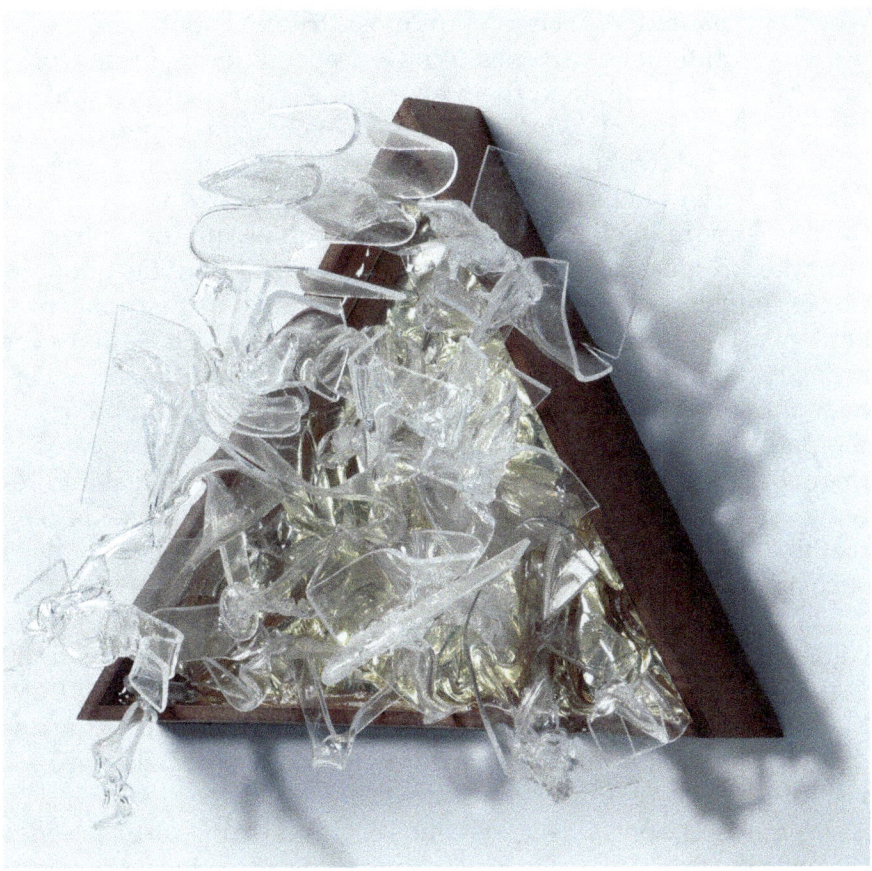

Figure 2.12. Student work, 2007. Plexiglas, mirror, epoxy, MDF, 14 × 14 × 6 in.

These are typical student works. Each was created through the manipulation of materials, and each relies almost completely on that process to carry the significance of the piece. At the outset of the project, students were not seeking to create a conceptually pre-determined work—they were not thinking about an idea; rather, they allowed the significance of the work to come forth through their choices and acts. In a sense, students once again "came upon" what they wanted. What is perhaps most striking about these pieces is their definitional elusiveness. Viewers encountering these works must come to understand them almost solely on their physical attributes, and students working in this fashion again experience the power of those attributes to affect viewers. Asked what these pieces "mean," we would be at a loss to answer. We are forced to approach them in a more intellectually mature way, allowing their physical attributes to speak to us. We sense in these works a subtle, nuanced presence that seems to belie the amateur status of their makers—student artists. Indeed, students working in this fashion—first-year college students—seem more adult, noticeably more mature. Perhaps the secret to that perceived maturity is the indirect nature of the path that they have followed. As has been noted previously, many beginning students are prone to direct message making, relying on easy symbolism to promote their direct statements. The method employed here, however, by its very nature discourages such an approach.

We now turn our attention to another work, this one employing colored felt (fig. 2.13). The student first cut the felt into small, irregularly shaped pieces and then sewed them together using thick black thread. The assembled structure was then attached to the underlying box, reiterating its shape, though now in a slightly askew fashion.

Reminiscent of stained glass, mosaics, and various artistic styles that employ the bold outlining of shapes, this attractive work might at first seem less inventive than the works shown in figures 2.11 and 2.12. But we must be cautious with such an assessment: The students who created the first two works approached the box as a container to be filled. Since students at the outset were told only to relate to this structure using manipulated materials, the first two students fell upon the most obvious solution in filling the box. This student, however, reacted quite differently, denying the identity of the box as a container. Seen from this perspective, his gesture becomes quite adventurous. This type of thinking—the breaking of expectations and assumptions—needs to be noticed and encouraged in the classroom, for it lies at the core of the inventive act.

In the next work (fig. 2.14), we see tubes of various lengths protruding from the base of the box. Each tube is topped with a clear

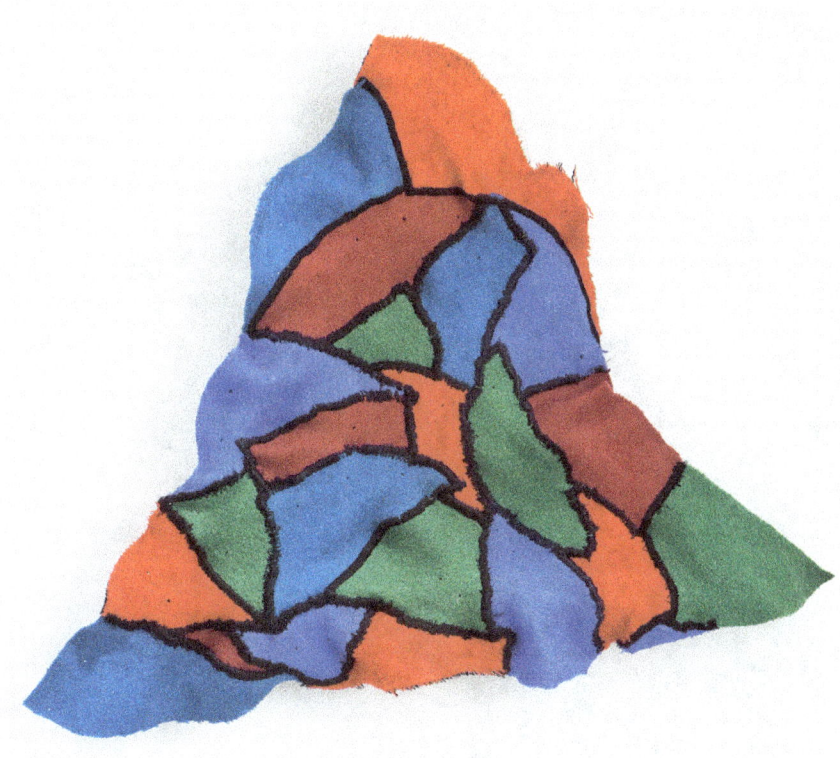

Figure 2.13. Student work, 2010. Felt, thread, MDF, 16 × 16 × 2 in.

plastic disc and a metal lock washer. The tubes are constructed of one-inch PVC pipe. The work is painted in a binary fashion—black and white. This simple and easily understood system of coloration, however, seems to be at odds with the placement and length of the tubes, both of which lack the "this and that" clarity of the color scheme. In place of order and predictability, we experience randomness, perhaps chance. Consequently the work, which was arrived at intuitively by the young artist, creates an interesting conversation.

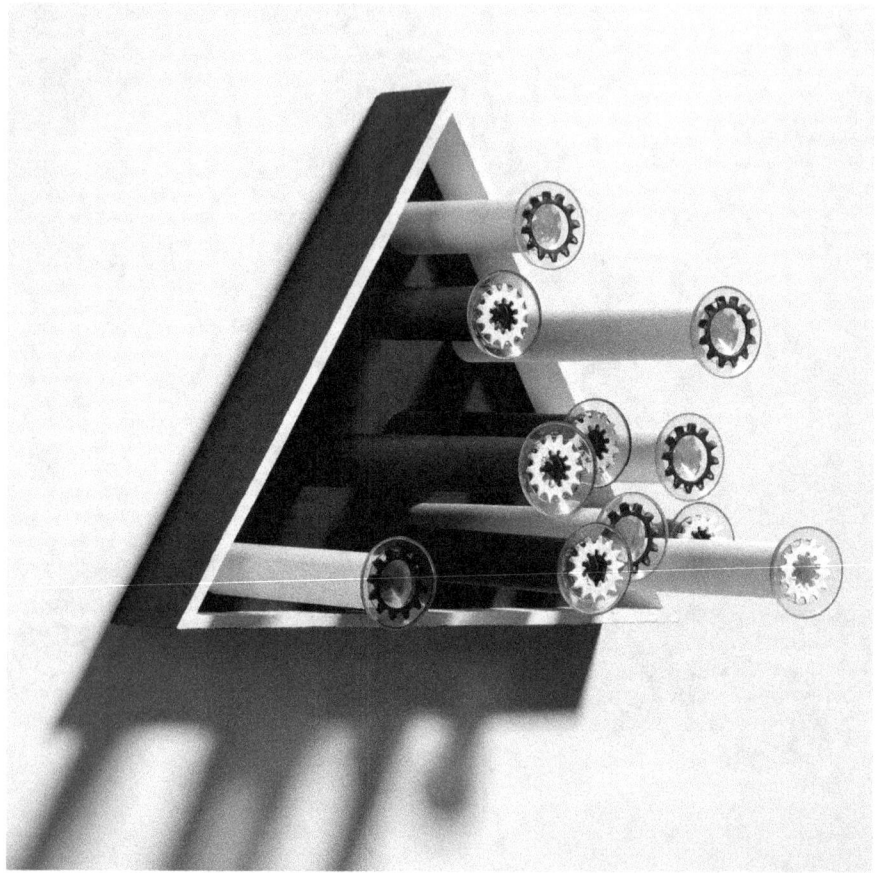

Figure 2.14. Student work, 2012. PVC pipe, plastic, metal washers, MDF, 12 × 12 × 10 in.

Works such as these help students gain artistic self-knowledge. And with this knowledge comes the possibility of change: decisions that are intuitively arrived at can be rationally scrutinized and altered if it is deemed necessary. It is important to structure both classroom discussions and mentoring conversations in such a way that students come to understand the logic of their artistic products so that decisions about change or development can be based on rational understanding. Only then can the teacher's advice stand on solid footing and evaluative decisions not seem arbitrary to students.

Many students enter the class with a humanistic bias—that is, they feel that art should speak directly about human concerns. The next work (fig. 2.15) was created by a student who started the course from this perspective. Early on, hoping to create a very significant piece, she fabricated a series of abstracted human hands by pouring liquid plaster into disposable gloves. But as the course unfolded and she experienced works of art that did not speak directly to human concerns, she came to question her artistic assumption. Faced now with the next assignment and the obligation to manipulate a material, she chose to break off the fingers of those hands, thus producing a large quantity of small, three-dimensional elliptical units. Though the units still retain some "finger" quality, the sense of "hand" has been lost through her act.

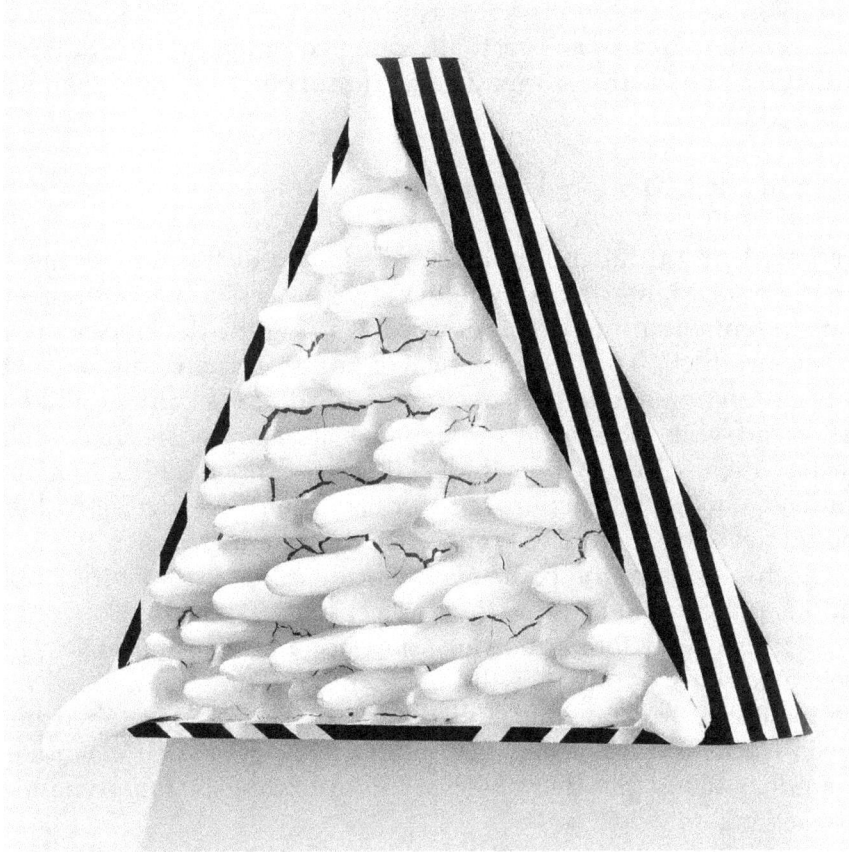

Figure 2.15. Student work, 2010. Plaster, glue, paint, MDF, 13 × 13 × 4 in.

This piece evolved over a three-week period. As the work was nearing completion, the student decided to cover the interior bottom surface of the box with glue. But she painted that surface before it was fully dry, and as a result, the surface cracked. At first discouraged, she soon realized that she liked the new surface. Perhaps she felt that this surface created an interesting dialogue with the alternating black-and-white stripes on the box. Perhaps she felt that the cracked surface added another layer of visual information, thus creating a more complex visual presence. For whatever reason, the new surface now felt right to her, and she accepted this chance occurrence. This choice, as well as others she made, indicates an evolving intuitive sense. The direct statement of "hand," so heavily laden with cultural significance, has given way to a maturing sensibility. Her quality of thinking has undergone a radical transformation, as the obvious and easy has given way to the implied and complex. Her story is actually quite common. Many students undergo a similar transformation as they experience the dynamics of the course.

Assignment 8: Mentored Works

We now turn our attention to the conclusion of the course: mentored works. Mentoring is a collaborative process in which teacher and student enter into discussions concerning the development of artistic products. The best mentor will not be intrusive but will act as a stimulating and supportive guide. Ideally, the student will be in the lead, with the teacher serving in an advisory position. For the teacher, the difficulty arises in deciding how much advice to give. Students should be encouraged to make their own decisions but should never be abandoned if they are running into difficulty.

At this stage of the course, each student, having manipulated various materials using a variety of techniques, has developed an artistic history, a foundation of knowledge and proficiencies. In their final work, students will build on this foundation, aided by the mentoring relationship and the trust that has developed between student and teacher. Let us now look at several examples of student work that demonstrate the unfolding of this dynamic interaction.

The first piece (fig. 2.16) presents a pair of highly finished wood blocks connected by curvilinear tubular elements made with epoxy resin. The piece is relatively small, measuring about thirty-six inches across, and rests securely on the floor. It was created by the same student whose triangular box with bent Plexiglas elements was discussed earlier (see fig. 2.12). In that work, the student used epoxy resin and became proficient with a heat gun. Material, tool, and technique show up again in this work: The tubular elements were created by pouring liquid epoxy resin into long, rigid plastic tubes that were then cut away, freeing the dried epoxy. The student then used a heat gun to bend the brittle epoxy to the desired shape. Equally important for the student was the intellectual understanding, gained earlier in the course, that a mature work of art can be created by the simple act of connecting elements. This piece also reinforced for the student the idea that conceptual significance follows, and is dependent on, the physical attributes of the work.

Besides being visually attractive, this work demonstrates both the subtlety and the complexity of thinking that we seek to develop in students. While being both economical and direct—two blocks connected by tubing—the work creates an intriguing mysteriousness, uniting quite easily the physical and the intellectual, the serious and the whimsical, in a work with no easy answer to the question, What's going on here? We are left with possibilities but no obvious answers.

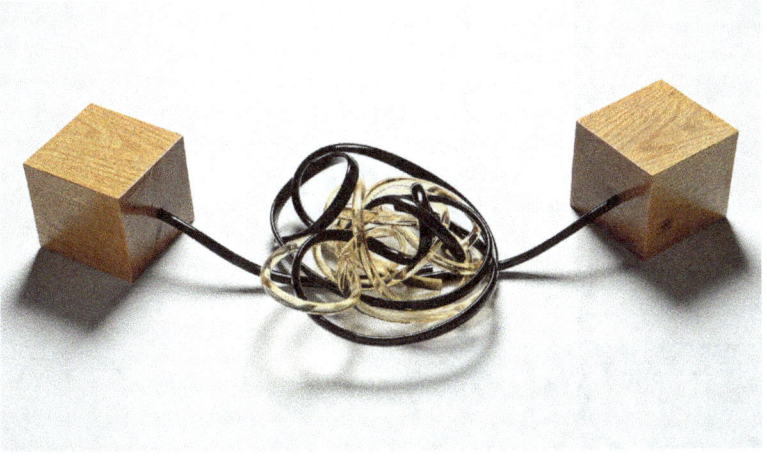

Figure 2.16. Student work, 2007. Wood, epoxy, 6 × 36 × 12 in.

We next turn our attention to a piece (fig. 2.17) measuring about eight and one-half feet in height and made primarily of steel. The height of this piece can easily be changed: the top portion screws down into a hollow base and allows for whatever adjustment is needed for a particular site. There is a removable central element placed at eye level, for which the artist created three separate units, each from different materials—card stock (fig. 2.18), toilet paper (fig. 2.19), and cloth and mesh (fig. 2.20).

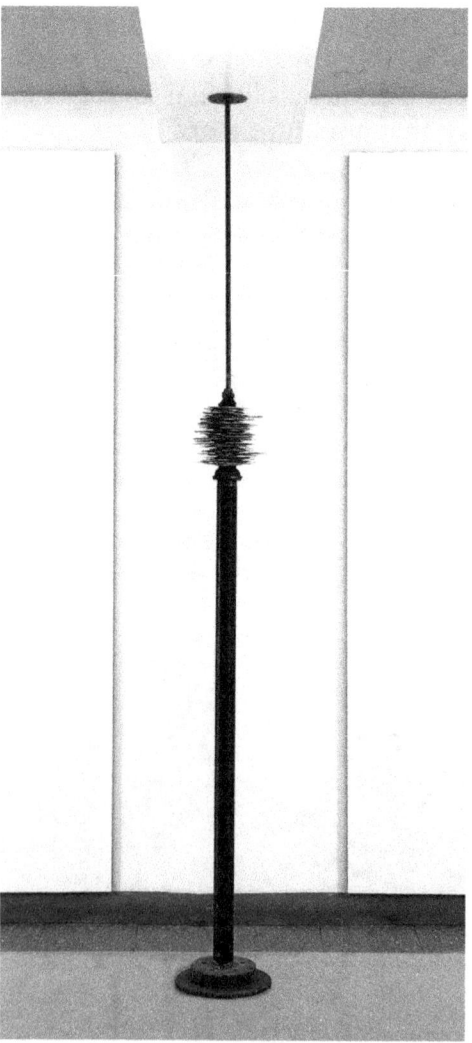

Figure 2.17. Student work, 2010. Steel, card stock, 8½ ft. (*as configured*).

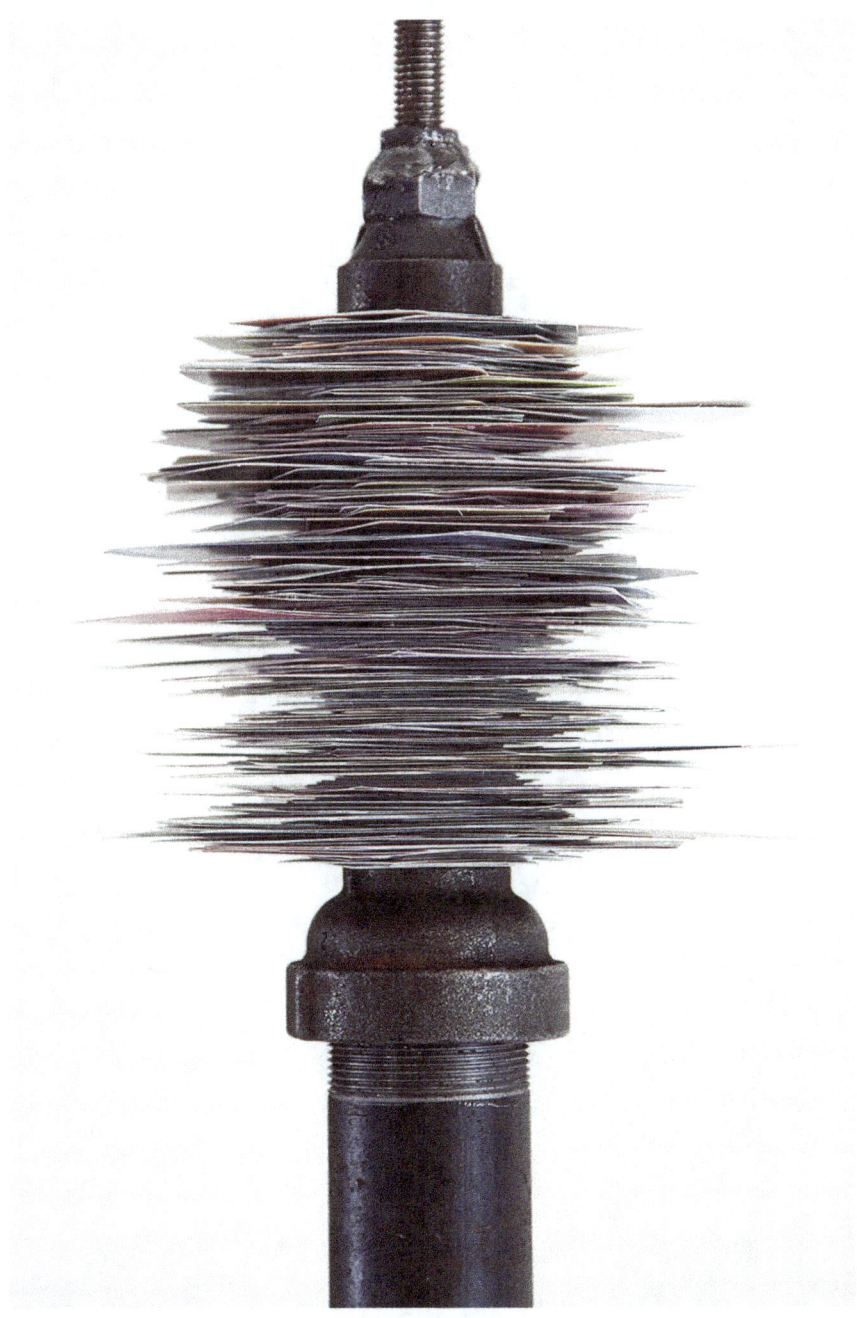

Figure 2.18. Student work (*detail*), 2010. Option no. 1, card stock.

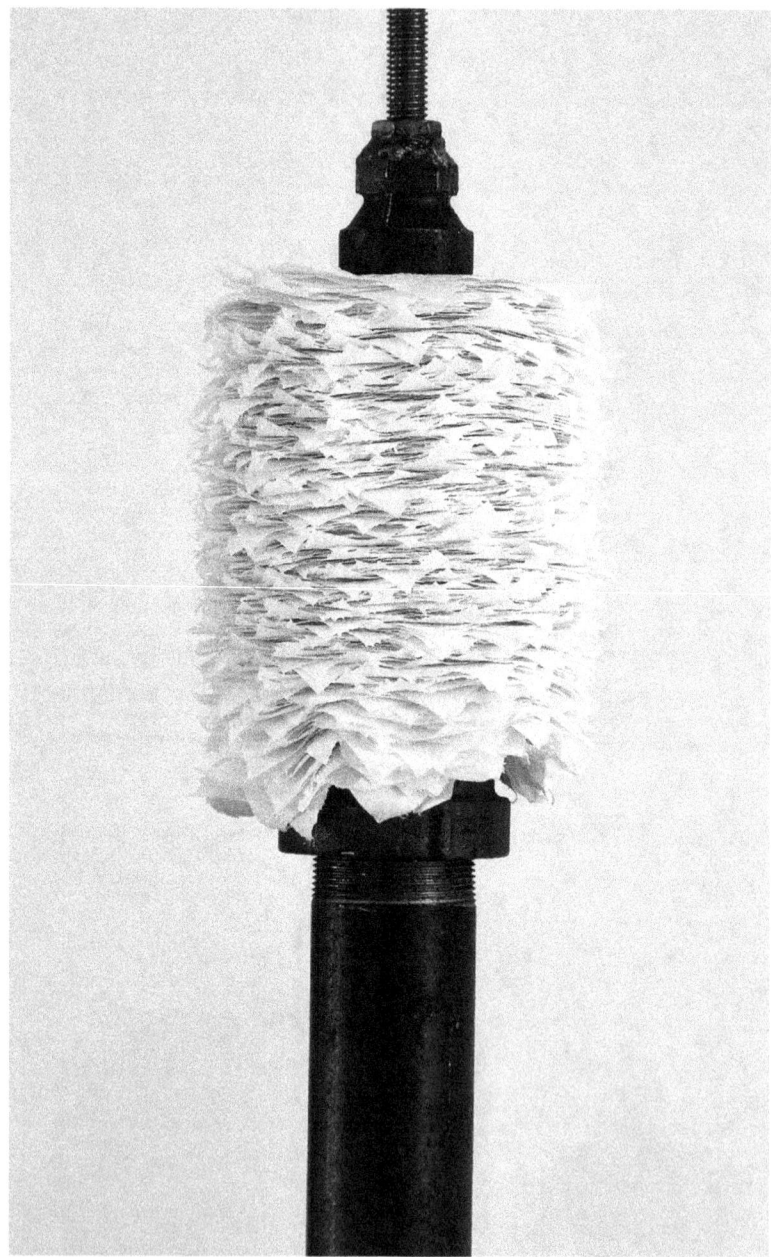

Figure 2.19. Student work (*detail*), 2010. Option no. 2, toilet paper.

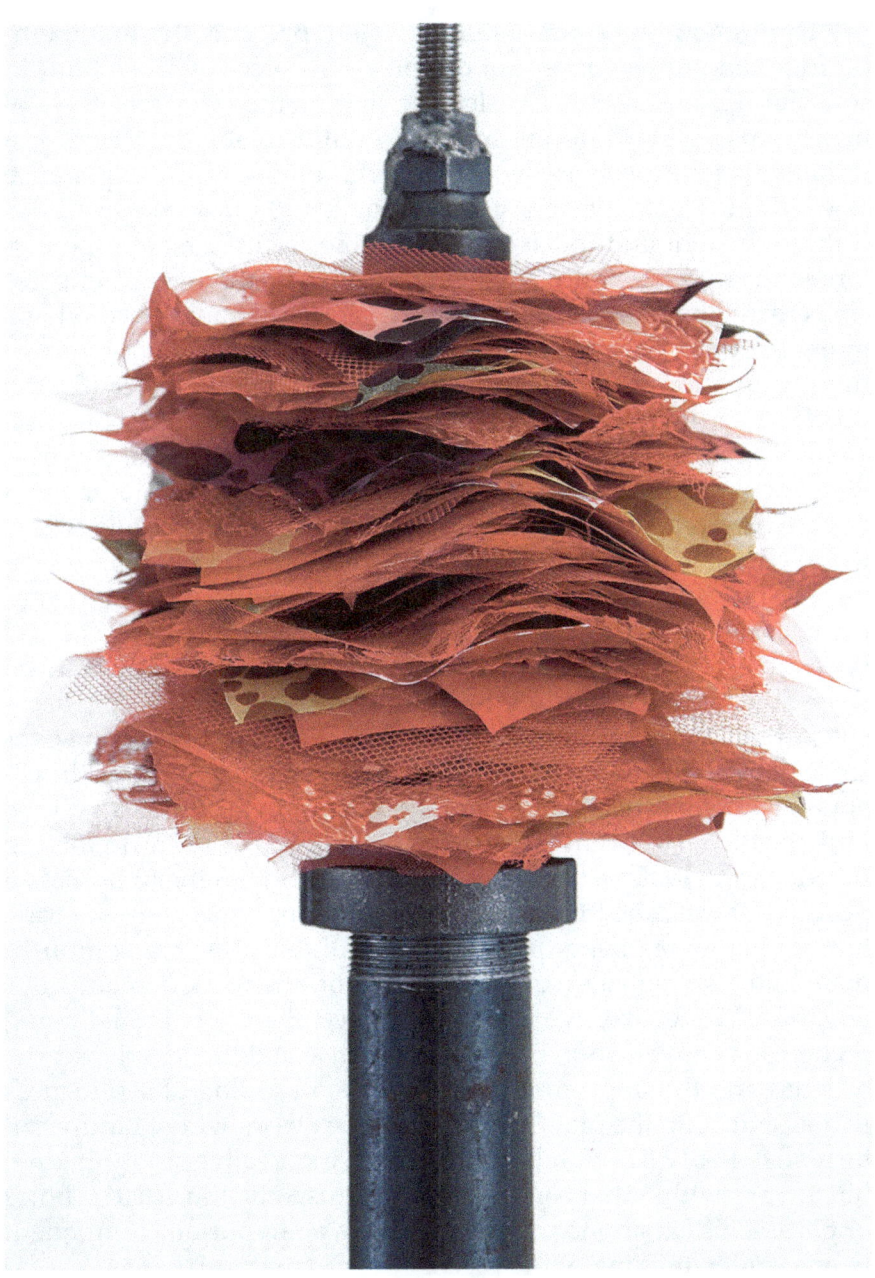

Figure 2.20. Student work (*detail*), 2010. Option no. 3, cloth, mesh.

Throughout the course, students were presented with a method founded on the concept of options. This piece, with its built-in series of options, seems to celebrate that method. Each of the elements, possessing differing material qualities, offers the viewer a unique visual experience, and usually a unique tactile experience as well, since most viewers seem compelled to touch the piece. But if through sight and touch a viewer comes to understand the materials in those central elements as "known," the overall impression created by this self-assured piece is one of strangeness. What's more, the artist has created a work with inherent potential for further exploration. Any number of materials could be placed in the central area. Imagine, now, long strips of cloth or pieces of glass. The intellectual ramifications of such a piece, a piece lacking a fixed identity, are extraordinary. The student has managed to create a work where potential is at least as important as presence and where the dynamic of options is as real as the options chosen.

The next work (fig. 2.21) has its origins in the first paper assignment and the act of curling—much as one does with gift ribbons. As the course developed, the material changed: we now see aluminum strips, but the act of curling has remained. Those strips, precisely cut and attached (fig. 2.22), now form a large piece, measuring over seven feet in height, constructed for a specific place—a backlit triangular opening between beam and ceiling, high up in a gallery, perhaps fifteen to twenty feet above the viewer (fig. 2.23). In forcing the viewer to look nearly straight up, the work seems to possess a peculiar power, one that causes physical pain. We must look away, but not before realizing the uniqueness of this artistic accomplishment. Sited so high above the gallery floor, the work demands our attention. We realize, even at that distance, that there is a lot going on here—a conversation between the dense and the less dense, the ordered and the impromptu, and an overall feeling of seriousness in dialogue with the playful. It is a complex work whose nature can be realized only through looking—studying, really—and thinking, but always within that context of pain, caused by the artist's choice of place. This experience of pain becomes an important component to any understanding of the piece.

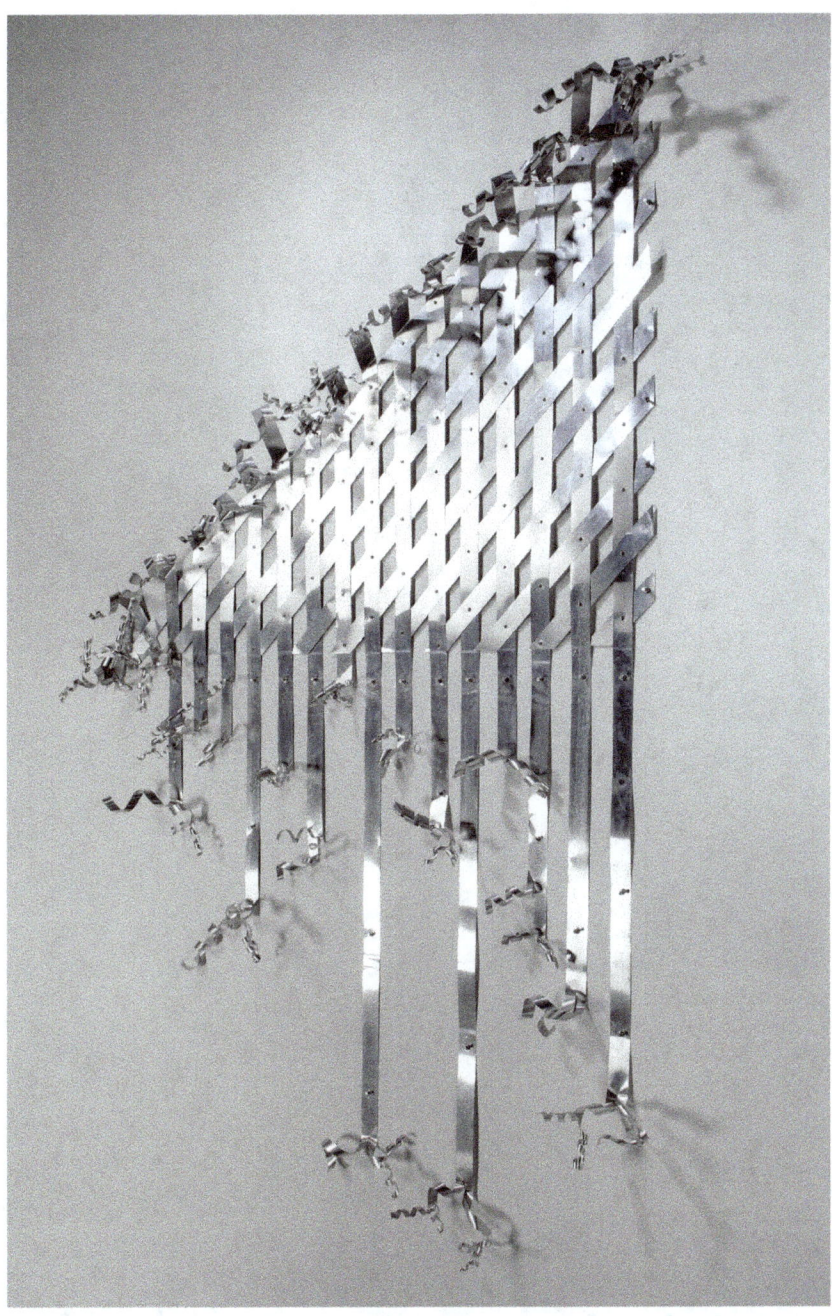

Figure 2.21. Student work, 2005. Aluminum, steel, 7 ft. × 4 ft. × 10 in.

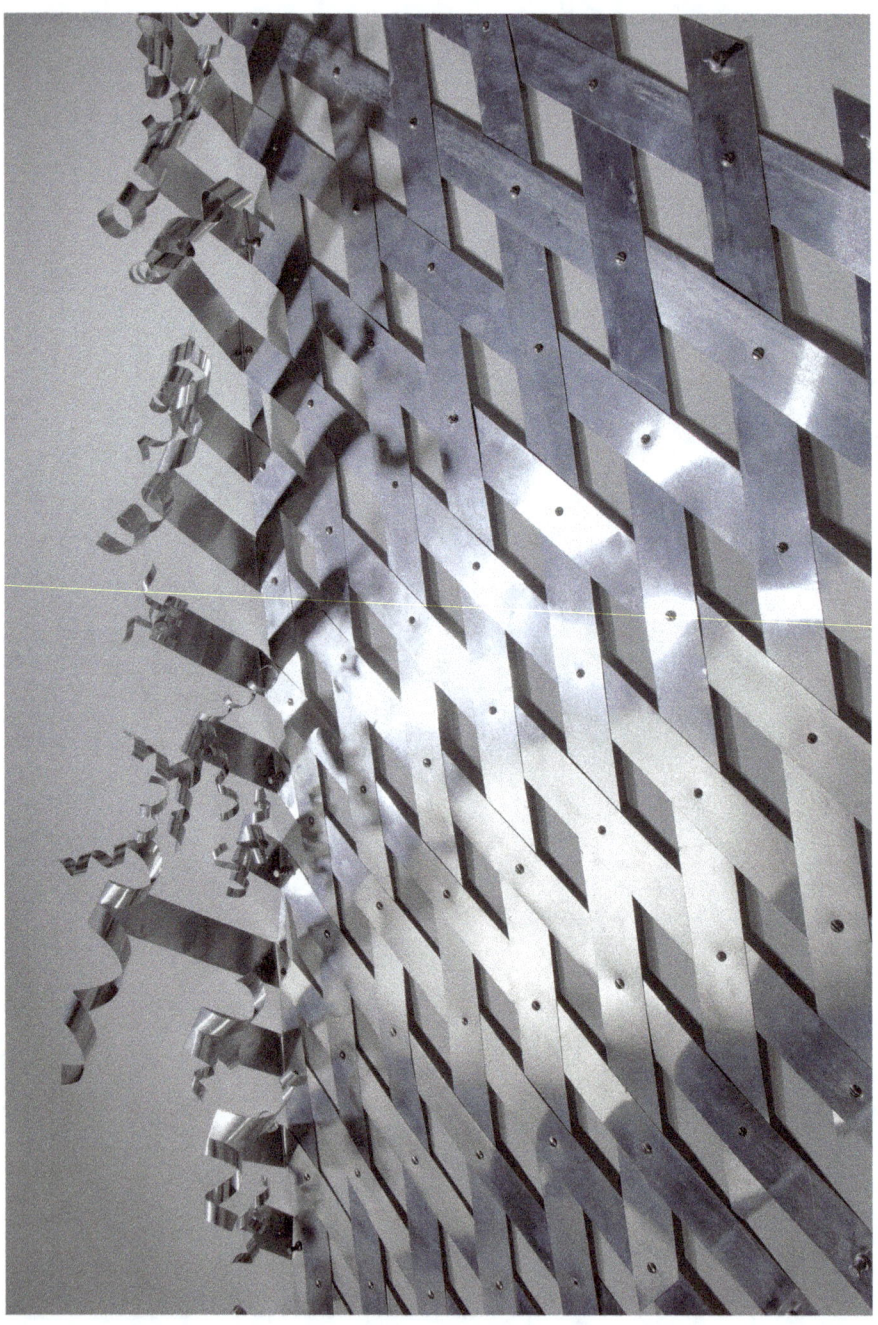

Figure 2.22. Student work (*detail, top edge*), 2005.

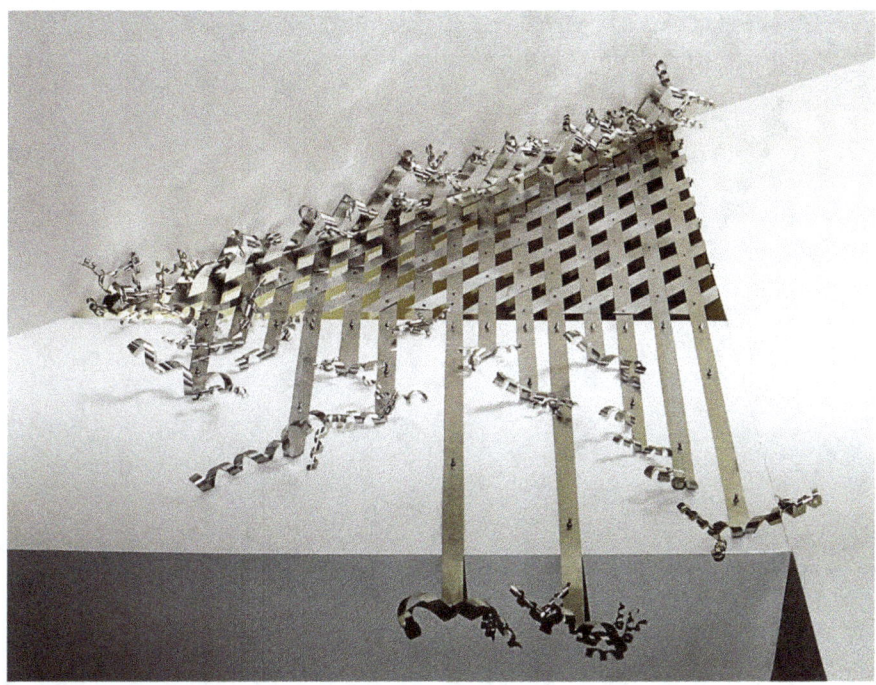

Figure 2.23. Student work (*installed*), 2005.

We end our discussion with a very large work, a work extend-
ing vertically about twelve feet up the wall (fig. 2.24). Like the previous work (fig. 2.21), this piece had its creative origins in the first
paper project. At that time, the student brought to class a roll of
heavyweight paper. Unrolling the paper seemed an obvious thing
to do. Laid out flat, long and narrow strips of paper soon became a
place of experimentation with paint and its application.

Using hollow plastic straws, the student carefully placed drops
of paint down on the paper's surface. Over time, the student experi-
mented, and soon she was placing drops of paint, at just the right
moment, on previously laid-down drops. Sometimes, drops were
extended, thus creating organic lengths of various shapes. Color
became a primary concern for the student, and a complex medley
of color relationships developed. Guided by intuition and demonstrating the necessary patience and will to accomplish the task, the
student moved slowly along the strips until finally satisfied with

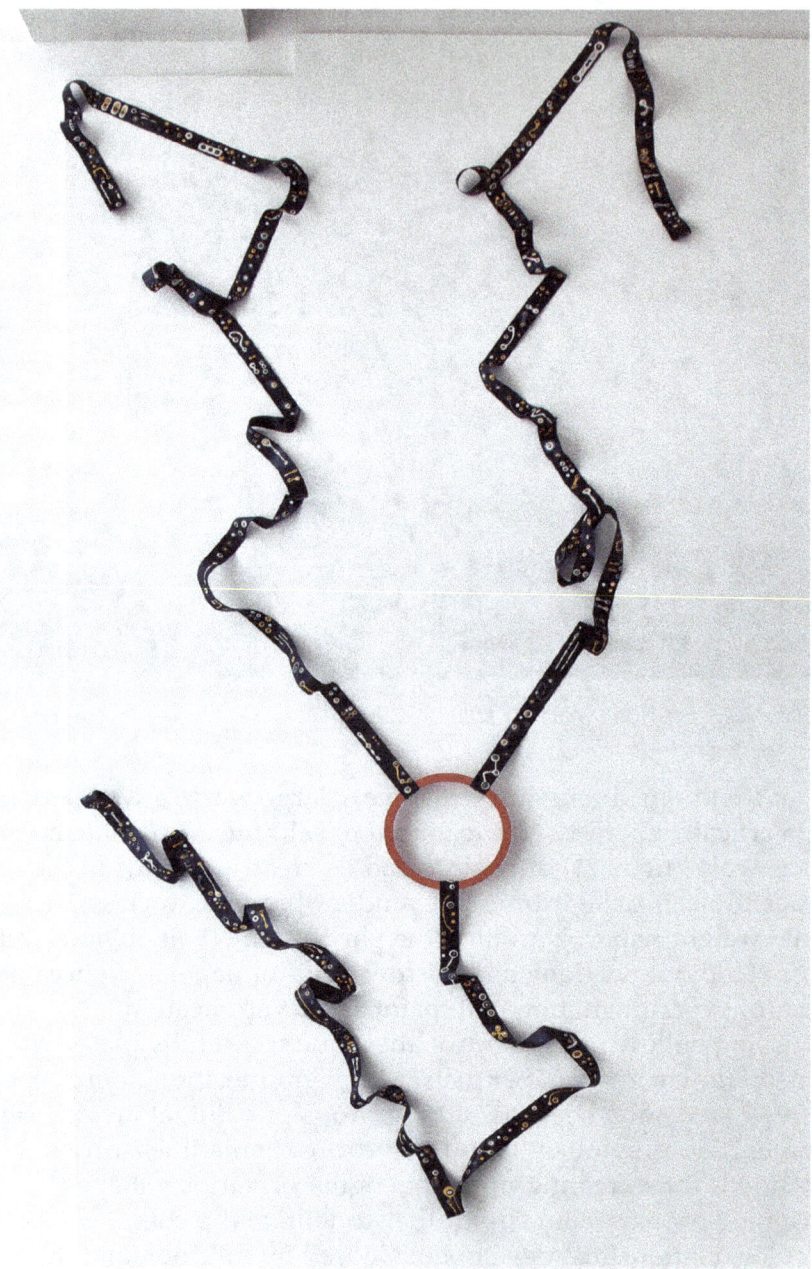

Figure 2.24. Student work, 2011. Paper, acrylic paint, MDF, 12 ft. × 6 ft. × 6 in.

the result (fig. 2.25). The final evolution of the piece, however, was still far off, since the question of display was still to be answered. The most obvious answer, the one that might at first have seemed intuitively right, was to place the strips flat on the wall, side by side. But guided by the desire for a new approach, the student searched for a novel solution and soon realized that those strips could be twisted and turned into any number of configurations. Perhaps the new curvilinear quality of the paper caused the student to think about that central unit in circular terms, or perhaps the chaotic freedom of the strips now needed a definite place to radiate from. For whatever reason, the student soon chose a raised circular element as a focal point.

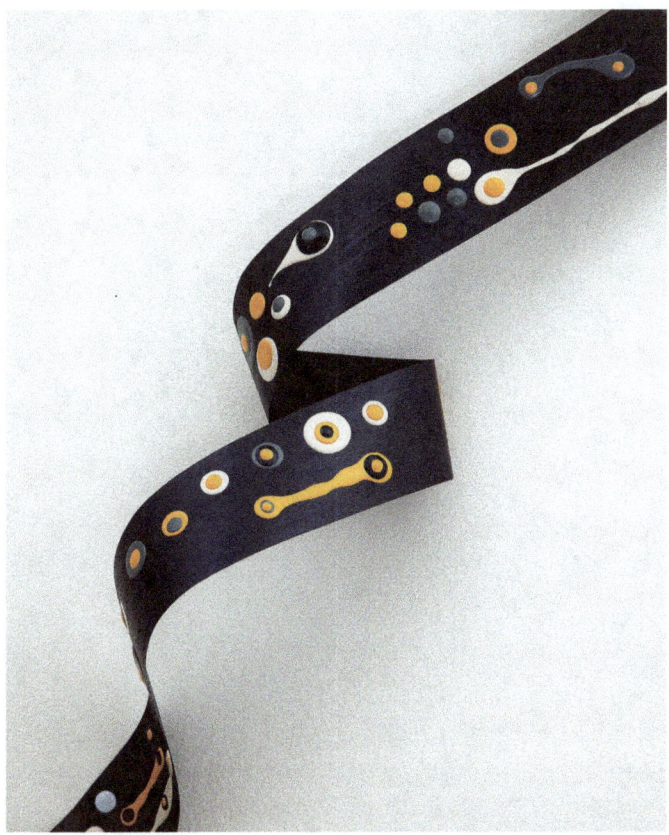

Figure 2.25. Student work (*detail*), 2011.

There is a physically overwhelming quality to this piece. Darkly defined on the white wall and towering over the viewer, the piece dominates the gallery space. At the same time, the piece demands intimacy as the viewer follows those endless drips and elongations. Power and intimacy create a work of great visual interest, demonstrating the unfolding of an artistic self.

Conclusion

The method described in this book accepts intuition as a creative force, but with limits. Too often an intuitive response is an easy response, providing only comfortable, familiar answers—answers that "seem right." By rationally guiding students' intuition with a structure of options, an ethos of exploration, and the goal of creating the unnamed, the teacher can enhance and redirect students' intuitive responses. Creating unexpected situations and posing questions that challenge assumptions can foster the breakdown of patterns and expectations and thus strengthen the foundation upon which students' intuitive decisions rest.

But the process can be destabilizing, and students often respond emotionally. Most students come to the class with a definite idea of what a work of art should look like; they usually come to the class having themselves created works that make a direct, often obvious statement, use representational imagery, and display strong emotion or conviction. The questioning of these oftentimes firmly held ideas about what art should be, and the breaking of the resulting patterns of creative production, can be emotionally upsetting, challenging students' sense of rightness, even at times their sense of self. Of course, we all organize our lives around patterns and live according to ideas. The teacher's task is to acknowledge the difficulty of what is being demanded and to mentor the student emotionally as well as creatively throughout the process, while never sacrificing the ultimate goal of inventive art making.

www.ingramcontent.com/pod-product-compliance
Lightning Source LLC
Chambersburg PA
CBHW070316290526
45791CB00003B/1128